On the Fringe of Society: Archaeological and Ethnoarchaeological Perspectives on Pastoral and Agricultural Societies

Edited by

Benjamin A. Saidel
Eveline J. van der Steen

BAR International Series 1657
2007

Published in 2016 by
BAR Publishing, Oxford

BAR International Series 1657

On the Fringe of Society: Archaeological and Ethnoarchaeological Perspectives on Pastoral and Agricultural Societies

ISBN 978 1 4073 0093 1

BAR Publishing is the trading name of British Archaeological Reports (Oxford) Ltd. British Archaeological Reports was first incorporated in 1974 to publish the BAR Series, International and British. In 1992 Hadrian Books Ltd became part of the BAR group. This volume was originally published by Archaeopress in conjunction with British Archaeological Reports (Oxford) Ltd / Hadrian Books Ltd, the Series principal publisher, in 2007. This present volume is published by BAR Publishing, 2016.

Printed in England

BAR
PUBLISHING

BAR titles are available from:

BAR Publishing
122 Banbury Rd, Oxford, OX2 7BP, UK
EMAIL info@barpublishing.com
PHONE +44 (0)1865 310431
FAX +44 (0)1865 316916
www.barpublishing.com

CONTENTS

On The Fringe Of Society: Archaeological And Ethnoarchaeological Perspectives On Pastoral And Agricultural Societies: Introduction

Eveline J. van der Steen and Benjamin A. Saidel

Pastoralism and farming: a brief history of research

The seeming dichotomy between farmers and herders (or pastoral nomads), agriculture and pastoralism, and, related to it, 'the desert and the sown' has been a major focus for archaeologists and historians of the Near East from the beginning of the discipline. It has also been intimately intertwined with another topic of research that involved not only historians, but also biblical scholars. [Specifically, many Old Testament scholars have interpreted Israel's Patriarchs, and also the people of Israel, as pastoral nomads, that is to say Bedouin *avant la lettre*. Apart from the Hebrew Bible, Egyptian sources, beginning in the Early Bronze Age and continuing into the Early Iron Age, refer to the Shasu and the Sutu, pastoralists who were also involved in raiding and robbing (e.g. Giveon 1971). Further north, the Mari archives describe conflicts between nomadic tribes and sedentary agriculture societies during the Old Babylonian period (Rowton 1973,1977 with references). Assyrian, Roman, Byzantine and Islamic sources mention the incursions of nomadic pastoralists and robbers, and Ibn Khaldun put the Arabs foremost in his description of human civilization, the first book of his Muqaddima, or introduction to history (Ibn Khaldun 1967: 97-116). On the other side of the line was the civilized world: the townspeople and the farmers, who were often the victims of the Bedouin, being raided and robbed, or extorted by them, especially when governments were weak or absent.

The written sources, which generally depict these wandering pastoral nomads depreciatingly as primitive, raiding and robbing groups (with the notable exception of the biblical patriarchs and the people of Israel), are contrasted by ethnographic models. Traditionally, many ethnographies underscored the ideals of Bedouin societies, although they also, like the historical sources, continued to present them as opposed to agricultural and urban societies (Evans-Pritchard's study of the Nuer in Somalia (1940) is a good example of this idealization).

These two misconceptions about pastoral nomads, their inherent nobility and their opposition to the settled community, are partly rooted in the 19th century, when travellers such as Burckhardt, Doughty and others adopted the bedouin lifestyle themselves, traversed the desert on camels, lived in the black tents, and wrote extensive accounts of those travels. They paint a picture of wild, roving bedouin, who detests the farmer and the settled land, and only makes incursions into it to rob the farmers of their grain (Burckhardt 1822, 1831: 1-18; Lynch 1849: 182; Doughty 1888; Palgrave 1865).

Later travellers have to a large extent confirmed this picture, such as Gertrude Bell (1907), and Wilfried Thesiger (1959; his introduction to Bailey 1991), whose sympathies generally lie with the bedouin, their free, roving lifestyle, rather than with the hard-working, poor, and somewhat pathetic farmers in the settled lands of northern Palestine and western Jordan. The Hebrew Bible has contributed to this picture with its accounts of the patriarchs of Israel Abraham, Isaac, Jacob, and the Children of Israel, roaming in the Sinai desert and through the lands of Edom and Moab.

Recent research, especially ethnographic but also archaeological, has considerably modified this picture (Layne 1994, numerous papers in Bar-Yosef and Khazanov 1992, Cribb 1991, Lancaster 1981; Salzman 1980, 2004; Marx 1967). Still, in most of these studies, pastoralists and agrarians/urbanists are seen as distinct societies, interacting, trading, competing, but separate.

Bar-Yosef and Khazanov, in the introduction of their 1992 volume state as one of the purposes of the volume: "to try to understand the intricate relationships between pastoral and agrarian-urban societies, based mainly on the archaeological records of the region".

The contribution of archaeology can be found in the way it has interpreted material culture, especially in marginal areas, that traditionally host nomadic pastoralists alongside settled farmers. Again, this picture is more complex than it is often painted, the major distortion being caused by the fact that nomadic pastoralists acquire much of their material culture through exchange with the sedentary societies (Musil 1928; Saidel 2000:211).

That there were contacts between the nomads and the settled world was recognized by all researchers. There were necessarily trade relations, mainly for foodstuffs produced in the different environments: meat and milk products in by the pastoralists, cereals and vegetables by the farmers. Another axioma was that part of the settled population would shift to a nomadic lifestyle when circumstances deteriorated, usually as a result of climatic deterioration, especially on the border region between the desert and the sown, or because of the weakening of central powers. On the other hand, a strong government, or climatic improvement, would induce pastoralists to settle down and become farmers (e.g. Hopkins 2003: 108). At the same time, the opposite has been suggested (Khazanov 1984; Finkelstein 1995: 38; Avni 1996), namely that specialized nomadism and pastoralism could only occur in a society with a strong sedentary or urban element. Köhler-Rollefson (1992) has developed a model that describes the operatives behind this specialization in the Neolithic period.

Another ambiguity in this concept of the divided society is the question of the definition of the word 'bedouin'. According to many modern scholars, and also to many early travellers, there were no 'real' bedouin in the Southern Levant. 'Real' bedouin were camel breeders, who would live in the desert with their camels for months on end, often the year round, and who would only come to the settled areas to either rob, or occasionally to buy grain. Sheep/goat pastoralists were no 'real' bedouin, or Arabs. This point of view was introduced by Ibn Khaldun, and it has been adopted by many later scholars. Ibn Khaldun, the great 14th century scholar has made his own classification of the different peoples that made up his world (1967: 91-116): he identified 'natural groups': the sedentary people were the inhabitants of cities and countries, who lived beyond the level of bare subsistence, and whose labour provided them with certain luxuries. Another natural group were the people who lived off agriculture and animal husbandry. These lived in the desert and the mountainous regions, the agriculturalists in villages and the herders of sheep and goats in tents. The third 'natural group' were the Arabs, who bred camels and lived far out in the desert.

The consequence of this 'division', which is followed by many early travellers, and by many modern scholars (Maeir *et al* 2003; Lewis 1987: 124; Gubser 1973: 14 note1; Rowton, *passim*; Cribb 1991, *passim*), is that there existed no 'real' pastoral nomads, at least not as a separate class, before the domestication of the camel which, according to general opinion, started around the end of the 2nd millennium BC (Köhler-Rollefson 1993).

Nevertheless, in the southern Levant, certainly in sub-recent history, several tribes have been labeled as 'bedouin', but eventually none of these lived entirely from the breeding of camels, and the name 'bedouin' referred to an idea and an ideal, rather than to a lifestyle.

The complexity and fluidity of the social and economic relationships between settled and nomadic lifestyles, is the subject of this volume.

The papers collected here are the precipitation of a seminar which was held at the Albright institute, Jerusalem, on June 3rd, 2004, bringing together a group of ethnographic and archaeological researchers working in Israel and Jordan.

The seminar focused on the interaction between all aspects of pastoralism and agriculture in the southern Levant, from the Bronze Age to the present. Given the importance of ethnographic research for the understanding of social organization, a number of papers focused on recent and subrecent pastoral and agricultural societies.Other papers approached specific historical periods, such as the Early Bronze IV, the Iron Age and the Roman and Byzantine period, from the perspective of an economically flexible society, where agriculture and pastoralism (or agro-pastoralism) were readily accessible economic alternatives. In these case studies changes in subsistence patterns depended on economic circumstances rather than on a fixed ideal or lifestyle (Salzman 1980).

Theoretical papers

This perspective has been worded most clearly by Emanuel Marx (this volume and references). Marx sketches a picture of bedouin society in Sinai and the Negev that reflects Salzman's 'Adaptation and Response' model, but actually goes much further. In Marx's view, the bedouin living in Sinai and the Negev are participants in a complex, economically diverse society, part of the wider network that includes the towns and cities in the region. They take part in the trade networks, and in the social and political networks of the region, in a mutually dependent economic relationship, but without losing their own, specific identity. Their identity is not determined by their means of subsistence, their economy, but by other, social and tribal networks. According to Marx, because the Sinai and Negev, the land bridge between Africa and Asia, have always been influenced and disrupted by major political incursions and disruptions, the recent political events (from 1967 onwards) can be seen as a reflection of historical events from a more remote past. Therefore the 'nomadic pastoralist' in the history and archaeology of the region cannot be seen as a separate group anymore, it is rather an aspect of the life of the tribal groups in the region. There is no divide between the desert and the sown, they are two sides of one economic and social unity.

The world of the Southern Levant, according to Marx, is an integrated society, in which the various groups are defined by claimed common descent rather than economic classes, interacting and mutually dependent.

The historical framework that has shaped this society, both within the larger world of the Southern Levant and within the smaller, localized communities, has been defined by LaBianca in his model of 'great and little traditions'. Great traditions are defined as 'universalized collective heritage and knowledge' frameworks, dividing the history of the region into social and cultural 'blocks'. Each of these traditions (LaBianca defines nine of them) has been the catalyst for one or more major developments that contributed to shaping the world of the Southern Levant: Egyptian and Mesopotamian great traditions revolutionized the organization of food production and integrated the food-producing community (farmers and pastoralists) in the economic systems of their respective empires. The Arabian great tradition created a 'desert culture' which was the catalyst for much of the 'bedouin' tradition that evolved in later periods. Of the next six traditions the Islamic, which has roots in both the Canaanite and the Arabian tradition, has at present the greatest impact on society. The Canaanite tradition brought the cultural and religious base to the Levant that is still its major driving force; the Arabian tradition is, to a large extent responsible for the norms and values of the Islamic tradition.

The 'little traditions' that Labianca defines as 'localized indigenous heritage and knowledge' include the long-term, practical solutions that people find for food production and the acquisition of their daily needs, and the enduring laws of social organization on the local community level. They include: mixed agro-pastoralism, low-risk water management, shared land rights, residential flexibility, tribalism, hospitality (as a means of communication and creating social bonds), and honour (as a means of social control). Within this framework societies and communities can continue to interact flexibly, taking part in all aspects of society at choice, without losing their own identity, interacting with, but not dependent on the 'great traditions'.

This interaction between the 'great' and 'little' traditions is particularly well illustrated in the paper by Benjamin Saidel and Abd Barakat, which investigates the transformation in function and symbolic value of a specific type of coin. These milled Spanish coins entered the region as part of the trade economy of the great tradition of the Ottoman Empire. Over time they became part of the folk tradition of the bedouin and peasant population of the Sinai and Negev, as symbols of fertility. This transformation seems to have originated in the motif on the reverse side of these specific coins: that of two pillars flanking an orb. The pillars became a symbol of fertility, and as a consequence the coins became incorporated in the local folk medicine, and used to enhance fertility in women.

Ethnohistoric and Ethnoarchaeological Research

Several contributions in this volume use ethnohistoric and ethnoarchaeological research to investigate the settlement and social organization of recent Bedouin populations. Van der Steen has looked at the social and economic organization of tribally organized communities, taking the Kerak Plateau as a case study. She makes a case for an integrated tribal polity, centered on the town but including the Plateau, and consisting of a number of mutually dependent tribes, with a hierarchical structure and a leading tribe, the Majali. The political centre was the town, but its internal organization reflected the tribal organization of the Plateau. Several other towns in the region, such as Salt and Jauf, reflected the same organization. The crux of the hypothesis is that the organizing power in a complex community is not the means of subsistence, but the tribal affiliation. Taking as a starting point the assumption that a tribal organization has always been indigenous to the region, the question is asked whether this organization is reflected in the archaeological record and can be recognized.

The same question has been hotly debated in the recent past in relation to pastoral nomadism in the Southern Levant (Finkelstein 1995: 23-26; Finkelstein and Perevolotsky 1990; Rosen 1992) and it is addressed in the papers of Uzi Avner and Yuval Yekutieli. Uzi Avner conducted a study of the Heywat, a tribe from the southern Negev, in order to determine if the Haywat leave recognizable material remains that can testify to their way of life. Avner describes the material culture of the Haywat as a precipitation of the mixed economy of the tribe: camp remains reflecting pastoralism, agricultural implements stored in caves, and utensils acquired through trade. Artistic expression is represented by sculpture, which is rare in present bedouin societies due to religious restrictions, but may have been less so in the past.

Avner's analysis of the existing ethnographic material describes the Haywat as fully integrated in the complex of interacting communities. Territorial and water rights had to be negotiated with other desert tribes, but at the same time were a source of income, as water could be sold to passing travellers, land could be rented out to the inhabitants of Aqaba, and the produce of the land could be sold if there was a surplus. The Heywat's territorial rights incorporated also the right to protect stretches of the Hajj route, and protection rights in general, a considerable source of income. The Haywat therefore had a mixed economy consisting of pastoralism (their main subsistence base), agriculture, protection, and trade and exchange with the surrounding communities. The range of their economic activities confirms and illustrates Marx's conclusions about the integration of bedouin in the larger community.

Yekutieli's paper also addresses the visibility of pastoral nomads in the archaeological record. He, however, expands upon this issue by including a discussion of the construction, use and the reasons for the abandonment of a structure locally known as Beit al-Malahi.

Yekutieli uses aerial photography, archaeology and ethnography, to study the Bedouin settlement pattern in a small research area north of Beersheba. His research clearly demonstrates the limits of what archaeology can and cannot do. This area was occupied by several tribes of the Tiyaha confederation, who used a combination of tents and stone structures for dwellings and storage. What stands out in this small region is the positive correlation between rainfall and topography on the one hand, and subsistence and dwelling types on the other. Areas characterized by rough topography and/or relatively low rainfall were used for grazing herd animals. Here tents functioned as the primary dwelling In contrast, those locations with high rainfall and smooth topography were used for agricultural activities. The primary type of architecture found in this area was structures constructed of clay and stone. Yekutieli's ethnographic investigation made clear that it was common for a family to live in a built structure and have agricultural fields in one area, and tents and herds in another, either at the same time, or alternating on a seasonal basis.

The role of religion and sanctuaries

Yekutieli's study also documents the change in function of a family compound from a domestic to a sacred space. In this research area there is one building, Beit al-Malahi, that served, and still serves, as a local shrine for the Bedouin. Local shrines are dedicated to the cult of a local saint, and used to be ubiquitous in the southern Levant (Abu Rabia, this volume with references).[1] Yekutieli describes that when it was still a domestic dwelling, a certain room in the house emanated signs of holiness, and seemed pervaded by a sacred presence, which is why it was eventually turned into a sanctuary. As a result the place now serves as a local pilgrimage place, in spite of the fact that no known saint is connected to it.

Apart from their obvious religious significance, shrines were a meeting place for members of different tribes and as such had a function in maintaining peaceful relations in the region, and creating and consolidating bonds and alliances. One category of sacred shrines originated in the tomb of a weli (a saint or holy person) who had played, or was believed to have played, a vital role in the history of a specific tribe, usually a patriarch or an early hero. His tomb would play a defining role in the territorial rights of the tribe. (Burckhardt 1831: 147,280; Stanley 1856: 54; Jennings-Bramley 1906: 26,33; Musil 1908/II: 34). The Heywat had an ancestral tribal sanctuary south of Aqaba (Avner, this volume). The second category of sacred shrines would refer to the specific powers of the buried saint, powers of protection or healing powers. These shrines were a meeting place for members of different tribes and as such had a function in maintaining peaceful relations in the region, and creating and consolidating bonds and alliances. Beit el-Malahi belonged to this latter type. The Nebi Musa sanctuary, described by Abu Rabia, also belongs to this category. It was originally built by Sultan Baibars, in the 13th century, and has, over the years, grown into one of the holiest shrines of Palestine, a gathering place for both farmers and bedouin. Specific feasts held at the shrine include the Nabi Musa celebrations, a time when large groups of people gather at the shrine. Bedouin take advantage of this gathering of people by having their sons circumcised, because a circumcision doctor is usually present. Exchange of news, goods and expertise takes place, and bonds are created or renewed. Thus the importance of the shrine in forging ties and bonds is vital.

Resource Management and choices

The ecological system, and the ecological restraints in which a society finds itself determine to a large extent whether a person, or a group of persons become or remain farmer, pastoralist, migrant worker or taxi driver. International political constraints are a second steering factor, as is amply demonstrated by the changes caused by the birth of the State of Israel, and the subsequent changes in the lifestyle of the bedouin of the Negev (Marx 1967). A third cause, and one that is directly related to, and itself heavily influenced by the first two factors, is the market economy. As Marx has stated (this volume): the bedouin of the desert are fully integrated participants in the market economy. The choices that they make are largely rational and a response to political, economic and ecological circumstances. For example, the Beni Sakhr would like to continue to raise camels if it was still a lucrative option. They, however, have reverted to agriculture, military service and sheep husbandry as these are currently more viable alternatives to camel pastoralism.

Conscious choices, based on economic possibilities led to the settlement of the Nabataeans in the course of the first century BC. The Nabataeans, known from the earliest written sources as a fiercely nomadic society, living off raiding and trade, developed into a powerful and very rich kingdom in southern Jordan and northern Arabia. The Nabataeans also settled in and cultivated the Negev desert on a previously unheard-of scale. Various theories have been put forward to account for this transformation. In this volume Tali Gini proposes to see the transformation from a nomadic to fully sedentarized lifestyle as a process of self-organization. Self-organization can be seen as a community's response to a disruption of a stable state, whereby the community, after a period of 'chaos' and fluctuation creates a new balance, in correspondence with and in response to the new circumstances. Thus a new 'stable state' is created, that may differ completely from the previous one. The 'stable state' in which the Nabataeans functioned as nomadic traders was disrupted with the cessation of trade through the Negev, and the take-over by the Romans. The Nabataeans became heavily involved in the trade and production of unguents, which vastly increased their wealth, but which also forced them to adopt a more sophisticated political and social organization, and a settled lifestyle.

That a rational and responsive attitude can have its drawbacks is demonstrated by Addison's paper on water management, or rather, the lack of it, among the Fayez of al-Qastal, an *ashira* of the Beni Sakhr. Pressure from the Jordanian government combined with years of drought, and other troubles that had reduced many of the camel herds, resulted in the sedentarization of the Beni Sakhr (Lewis 1987: 127, 133-34). In Qastal, induced by

[1] Local shrines had a cohesive function in the tribal society of the Levant, as much as a religious one. This may explain why the Wahabis never fully managed to suppress them, in spite of the fact that they regarded them as pagan and destroyed many of them in the beginning of the 19th century. In the beginning of the 20th century they were still more or less ubiquitous (Schumacher mentions 27 separate welis in the Hauran and Jaulan: Schumacher 1886; 1888).

government subsidies on farming and imported water, the Beni Sakhr moved away from a pastoral economy to one based on agriculture. The Jordanian government's strategy was induced by pressure from the international market. (Walker, this volume). Eventually this resulted in the Fayez and other *asha'ir* turning into 'settled bedouin', who make their income through farming, but at the same time adhere to what they see as their 'true' bedouin identity. Exactly how they express this identity is described by Addison: stressing the importance of 'true' bedouin roots in their external relationships; occasional and temporary return to desert life in the black tent (usually only for a holiday), prowess with (fire) arms, and a show of wealth and hospitality, displayed in a luxuriant and wasteful use of water. Thus the Fayez, and others like them have made pragmatic choices based on the assets that were offered them. If these choices turn out to have disastrous effects in the long run, it is largely the government that is to blame, because both individuals and communities must be expected to choose what gives them the best returns on the short term.

Bethany Walker arrives at the same conclusion in her analysis of the Mamluk and Ottoman periods in two villages in Jordan. Traditional explanations for the decline in settlements during the Ottoman period focuses on climatic changes and neglect by the government. These factors are believed to have resulted in clashes between villagers and Bedouin, which ultimately led to the abandonment of many villages. In contrast to this traditional interpretation Walker prefers to see the general decline in settlement as the result of short-term responses to the market economy, both by the government and by the village farmers themselves. Both in the Mamluk and the Ottoman periods the State's policy was to maximize export products such as sugar and olive oil. Over time plots of land were reallocated by the government, in order to gain more control, either direct or indirect, over production. As these plots were used to supply the international market, they lay generally along transportation routes. This policy focusing on short-term gain proved disastrous in the long term. The more remote areas, where population was denser, and farmers generally provided for the local market, survived the economic crisis of the 15th century much better than the regions that were serving the international market. In the 19th century the Ottoman government promoted farming almost aggressively, in an effort to profit from the international market economy. These political and economic issues influenced the cycles of settlement and abatement at least as much as the traditionally proposed explanations, which focus on climatic changes and bedouin incursions.

Farmers or Pastoral Nomads

Five papers in this volume are dedicated to the interaction between settlement and pastoral nomadism in the Negev desert and the Wadi Arabah in various historical periods. Settlement patterns in the Negev have been the focus of much attention from archaeologists, starting with Nelson Glueck and his surveys of the Negev (Glueck 1959, 1960, 1965), and continuing to the present day. Much of this research focuses on fluctuations in the settlement patterns over time (Rosen 1987; 1993; Finkelstein 1995; Haiman 2003, with references; Herzog 1994) and on the question why people would choose to settle in an arid region such as the Negev, and consequently the interaction between farmers and pastoralists in the region, particularly in periods of settlement peaks (Haiman 2002, 2003; Rosen 2002a,b, 1993; Finkelstein 1995; Cohen and Cohen-Amin 2004). The central and southern Negev seemed perfectly suited for pastoralists, who were mobile and could seek out the best pasture, and whose needs in terms of water were relatively low. But the presence of settlement systems, within the relatively harsh ecological system of the Negev highlands has induced scholars to seek a link with the northern areas, on the assumption that the low rainfall and high temperatures made independent existence in the region hard, if not impossible (Rosen 2000, 2002a). Other explanations and hypotheses vary from climatic changes (Herzog 1994) and border protection (Haiman 2003) to independent trade networks (Finkelstein 1995,103 ff; Singer-Avitz 1999 Bienkowski and van der Steen 2001).

Tali Gini's paper on the Nabataean settlement in the Negev has been discussed above.

The paper by Vardi *et al* in this volume suggests that already in the Early Bronze IV period a trade oriented economy must have existed, which involved a relatively large area, including regions north and east of the Negev Highlands (also Cohen and Dever 1978, 1980, 1981; Haiman 1996). This conclusion is based on their analysis of the lithic assemblages from the Early Bronze Age IV (also EB-MB, Intermediate Bronze, or Middle Bronze I) settlements of Beer Risisim and Ein Ziq . Canaanean blades, very common further north, are absent on these sites. The limits of distribution seem to lay at Jebel Qa'aqir, south of Hebron. In the previous periods of occupation, the Early Bronze I and II, these limits lay further south, at Arad, but still north of the sites investigated here. This void in the toolkit was not reflected in the accompanying artifact repertoire, which suggested extensive contacts with the areas of Judah and the Shephelah, as well as the Central Jordanian Plateau. There was a cottage-industry production of beads, made from imported materials such as Red Sea shells. At the same time, the absence of Canaanean sickle blades for grain harvesting does not reflect a lack of cereal consumption, as shown by the numerous grinding stones that were found. Grain must therefore have been imported. The conclusion must be that Beer Risisim and Ein Ziq were part of a large, integrated network that extended over different ecological zones. The inhabitants could afford not to produce grain themselves, because they imported it, in exchange for animal products, and jewelry made from Red Sea shell and other prestigious materials.

Another settlement peak that has attracted much attention occurred in the Iron Age II (Aharoni 1967;

Cohen 1979; Eitam 1988; Finkelstein 1984). In the present volume, Haiman defends his position that the IA II settlements in the Negev highlands received economic support from the kingdom of Judah in exchange for protecting the southern borders of this kingdom. Haiman points out that for most of the sites there is no clear correlation between settlement location and either water sources or agricultural soil, and therefore the incentive to build and settle must have been external, directed by a larger polity. Pastoralism and agriculture existed side by side in a mixed economy that was adapted to the local circumstances, but still could only function in the context of a larger polity, in this case the kingdom of Judah.

This same group of sites, which has been interpreted by Haiman as part of an integrated political system, has been studied by Hendrik Bruins. His approach is a different one, focusing on the archaeological remains, and in particular the organic remains of the agricultural fields of one of these sites, Horvat Haluqim. Horvat Haluqim is a site in the arid zones of the Negev highlands, a region that is considered suitable for pastoralism. It consisted of a fortress, dated to the Iron Age II, and additional buildings. Its location, in the direct vicinity of three small wadis, made agriculture in this area feasable, through the application of terracing and runoff water catchment. One of the terraced fields, field 12, contained an anthropogenic soil layer, which was dated to the Iron Age II by means of dateable pottery and carbon-14 dating. Micromorphological analysis revealed that the inhabitants of the site used animal dung and household waste as fertilizer. It also revealed, through the presence of ferric nodules, control of the runoff flows used for irrigation.[2] Based on his micromorphological analysis Bruins proposes that the deposition of material culture and organic remains in agricultural plots is evidence of farmers manuring their fields. Both Haiman and Bruins agree in their conclusions that the 'fortress sites' in the Negev highlands do not represent a nomadic population in the process of settling down. They must rather be seen as experienced farmers, practicing a mixed economy of agriculture and sheep/goat pastoralism.

The next settlement surge in the Negev occurred at the end of the Byzantine and the beginning of the Umayyad period. Again, various theories have been put forward to explain this sudden surge of settlements, which are generally interpreted as settlements from sedentarizing nomads. Haiman sees them as a purposive policy by the Umayyad government as a buffer against nomadic incursions (Haiman 1995), while Avni (1996) adheres to the theory that nomadic pastoralists can only specialize in an urban environment, and interprets the settlements as the pastoralists' response to the decline of the Negev towns.

[2] The same phenomenon has been recognized at other sites; e.g. most recently Baumgarten 2004, 13.

Dov Nahlieli (this volume) puts forward a new model, in which he identifies the inhabitants of the new, small settlements as former inhabitants of the towns. With the decline of those towns, due to a combination of plague, earthquake and deteriorating climate, they abandoned their original hometowns and reverted to a rural lifestyle on farmsteads close to or directly on their agricultural fields. This was possible thanks to the protection and support offered by the Byzantine and later the Umayyad authorities. Once the seat of government moved to Baghdad, and government interest in the region waned, the settlements could not support themselves anymore, and declined.

Conclusion

In spite of their widely varying attitudes towards the specific phenomena and problems in the history and archaeology of the Southern Levant, all these papers make clear that agriculture, pastoralism, trade, exchange and other forms of subsistence were pursued within the same communities, either together, as a mixed form of subsistence, or subsequently, when political, economic or ecological circumstances made a change of subsistence viable. The desert and the sown are not divided, they are part of the same landscape, shared by the same community, and the borders lie somewhere else. Communities are not defined by the way they make their living, even though they may make different choices under seemingly identical circumstances. We, as archaeologists, as western scholars, have only a fragmentary view of the way ancient communities in the Southern Levant experienced their world. Their economic choices, which is what we are concerned with here, were determined by the way they perceived the world they lived in: the community to which they belonged, the communities that surrounded them, and their physical world: the soil, the climate and the landscape, both physical and ideational or even religious. In order to understand the choices these people made, we have to make an effort to understand their world (Maeir *et al* 2003; Bienkowski in press).

This is where ethnography, and ethnoarchaeology come in. Recently and subrecently there has been heavy pressure on the political, social and economic structures of the different societies in the southern Levant. These pressures have led to adaptations in economic pursuits, in political choices and in the social organization of the individual groups that form the social landscape of the region. As Marx points out in this volume, international pressures are nothing new in the region, and are bound to have forced people into making similar choices over time. At the same time, the basic social organization of the region has been stable over a long time period, being composed of tribal groups that interacted in an uneasy and fluctuating balance of antagonism and alliance.

Therefore, even though the recent challenges to society may seem more far-reaching than anything that has happened before, the way people have responded to these changes may not be so different from the way they responded to changes in the past, be it the Byzantine period, Iron Age or Early Bronze Age.

Bibliography

Aharoni, Y.
1967 "Iron Age fortresses in the Negev". *Israel Exploration Journal* 17: 1-17; 97-111.
Avni, G.
1996 *Nomads, Farmers and Town-Dwellers: Pastoralist-Sedentist Interaction in the Negev Highlands. Sixth – Eight Centuries CE.* Publications of the Archaeological Survey of Israel; supplement. Jerusalem: Israel Antiquities Authority.
Bailey, C.
1991 *Bedouin poetry from Sinai and the Negev. Mirror of a culture.* Oxford: Clarendon.
Bar-Yosef, O., and Khazanov, A., eds.
1992 *Pastoralism in the Levant.* Madison Wisc: Prehistory Press.
Bell, G.M.L.
1907 *The desert and the sown.* London: William Heinemann.
Bienkowski, P.
In press "The Wadi Arabah: Meanings in a Contested Landscape" in *Crossing the Rift: Resources, Routes, Settlement Patterns and Interaction in the Wadi Arabah*, eds, P. Bienkowski and K. Galor.
Bienkowski P., and van der Steen, E.J.
2001 "Tribes, Trade and Towns: a new framework for the Late Iron Age in Southern Jordan and the Negev". *Bulletin of the American Schools of Oriental Research* 323: 21-47.
Burckhardt, J.L.
1822 *Travels in Syria and the Holy Land.* London: J.Murray.
1831 *Notes on the Bedouins and Wahabis, collected during his travels in the East.* London: Colburn & Bentley.
Cohen, R.
1979 "The Iron Age fortresses in the central Negev". *Bulletin of the American Schools of Oriental Research* 236: 63-75.
Cohen, R., and Cohen-Amin, R.
2004 *Ancient settlement in the Negev highlands.* Israel Antiquities Authority reports no.20
Cohen, R., and Dever, W. G.
1978 "Preliminary Report of the Pilot Season of the Central Negev Highlands Project". *Bulletin of the American Schools of Oriental Research* 232: 29-45.
1979 "Preliminary Report of the Second Season of the Negev Highlands Project". *Bulletin of the American Schools of Oriental Research* 236: 41-60.
1981 "Preliminary Report of the Third and Final Season of the Central Negev Highlands Project". *Bulletin of the American Schools of Oriental Research* 243: 57-77.
Cribb, R.
1991 *Nomads in archaeology.* Cambridge: Cambridge University.
Doughty, C.M.
1888 *Travels in Arabia Deserta.* Cambridge: Cambridge University.
Eitam, D.
1988 "The settlement of nomadic tribes in the Negeb highlands during the 11th century BC" *in Society and Economy in the Eastern Mediterranean (c.1500-1000 B.C.)*, eds. M. Heltzer and E. Lipinski. Leuven: Peeters. pp 313-340.
Evans-Pritchard, E.E.
1940 *The Nuer: a description of the modes of livelihood and political institutions of a Nilotic people.* Oxford: Clarendon Press.
Finkelstein, I.
1984 "The Iron Age 'Fortresses' of the Negev Highlands: Sedentarization of the Nomads". *Tel-Aviv* 11:189-209.
1995 *Living on the fringe: the archaeology and History of the Negev, Sinai and neighbouring regions in the Bronze and Iron Ages.* Sheffield: Sheffield Academic Press
Finkelstein, I., and Perevolotsky, A.
1990 "Processes of Sedentarization and Nomadization in the History of Sinai and the Negev*". Bulletin of the American Schools of Oriental Research* 279: 67-88.
Giveon, R.
1971 *Les Bédouins Shasu des documents égyptiens.* Leiden: Brill.
Glueck, N.
1959 "An aerial reconnaissance of the Negev". *Bulletin of the American Schools of Oriental Research 155*: 2-13.
1960 "Archaeological exploration of the Negev in 1959". *Bulletin of the American Schools of Oriental Research* 159: 3-14.
1965 "Further explorations in the Negev". *Bulletin of the American Schools of Oriental Research* 179: 6-29.
Gubser, P.
1973 *Politics and Change in el-Kerak, Jordan.* London: Oxford University.
Haiman, M.
1995 "Agriculture and Nomad-State relations in the Negev Desert in the Byzantine and early Islamic periods". *Bulletin of the American Schools of Oriental Research* 297: 29-53.
1996 "Early Bronze Age IV Settlement Pattern of the Negev and Sinai Deserts: View from Small Marginal Temporary Sites". *Bulletin of the American Schools of Oriental Research* 303: 1-33.
2002 *Enclosed structures, pens and pastoralism in the Negev and Sinai deserts – past and present.* Toronto: American Schools of Oriental Research Meeting 2002.

2003 "The 10th Century B.C. Settlement of the Negev Highlands and Iron Age Rural Palestine", in *The Rural Landscape of Ancient Israel*, eds. S. Dar; A. Maeir; and Z. Safrai. British Archaeological Reports, International Series 1121, Oxford. pp 71-90.

Herzog, Z
1994 "The Beer-Sheba valley: from nomadism to monarchy" in *From Nomadism to Monarchy, Archaeological and Historical aspects of Early Israel*, eds. I. Finkelstein and N. Na'aman. Jerusalem: Israel Exploration Society. pp 122-149.

Hopkins, D.C.
2003 "Agriculture" in *Near Eastern Archaeology; a reader,* ed. S. Richard. Winona Lake: Eisenbrauns. pp 124-130.

Ibn Khaldun, A.
1967 *The Muqaddimah. An introduction to history. Translated from the Arabic by Franz Rosenthal.* Edited and abridged by N.J. Dawood. Princeton: Bollingen Series.

Jennings-Bramley, W.E.
1906 "The Bedouin of the Sinaitic peninsula". *Palestine Exploration Fund Quarterly Statement* 23-33,103-109, 197-205, 250-258.

Khazanov, A.
1984 *Nomads and the Outside World.* Cambridge: Cambridge University.

Köhler-Rollefson, I.
1992 "A model for the development of nomadic pastoralism on the Transjordanian Plateau" in *Pastoralism in the Levant*, eds. O. Bar-Yosef and A. Khazanov. Madison Wis: Prehistory Press. pp 11-28.
1993 "Camels and camel pastoralism in Arabia". *Biblical Archaeologist* 56:180-188.

LaBianca, Ø. S.
(in press) "Great and Little Traditions: A Framework for Studying Cultural Interaction through the Ages in Jordan" in *Studies in the History and Archaeology of Jordan No. 9.* Amman: Department of Antiquities.

Lancaster, W.
1981 *The Rwala beduin today.* Cambridge: Cambridge University.

Layne, L.
1994 *Home and Homeland. The dialogics of tribal and national Identities in Jordan.* Princeton: Princeton University.

Lewis, N.N.
1987 *Nomads and settlers in Syria and Jordan, 1800-1980.* Cambridge: Cambridge University.

Lynch, W.F.
1849 *Narrative of the US expedition to the river Jordan and the Dead Sea.* Philadelphia: Lea and Blanchard.

Maeir, A.; Dar, S.; and Safrai, Z.
2003 "Introduction: the rural landscape of ancient Israel" in *The Rural Landscape of Ancient Israel*, eds. S.Dar; A. Maeir; and Z. Safrai. British Archaeological Reports, International Series 1121, Oxford. pp ii-vi.

Marx, E.
1967 *Bedouin of the Negev.* Manchester: Manchester University.

Musil, A.
1907-8 *Arabia Petraea.* Wien: Alfred Holder.
1928 *The manners and customs of the Rwala bedouin.* The American Geographical Society.

Palgrave, W.G.
1865 *Narrative of a year's journey through Central and Eastern Arabia (1862-63).* London: Macmillan.

Rosen, S.A.
1987 "Byzantine nomadism in the Negev: results from the Emergency survey". *Journal of Field Archaeology* 14:29-42
1992 "Nomads in Archaeology: a response to Finkelstein and Perevolotsky". *Bulletin of the American Schools of Oriental Research* 287: 75-85.
1993 "A Roman-period pastoral tent camp in the Negev, Israel". *Journal of Field Archaeology* 20: 441-451
2002a "An economic model for Early Bronze Age pastoral nomadism" in *Beer-Sheba XV, Aharon Kempinsky Memorial Volume*, eds. S. Ahituv and E. Oren. Beersheba: Ben-Gurion University. pp 344-59.
2002b "The evolution of pastoral nomadic systems in the southern Levantine periphery" in *In quest of ancient settlements and landscapes*, eds. E. C. M. van den Brink and E. Yannai. Tel Aviv: Ramot, Tel Aviv University. pp 23-44.

Rowton, M.B.
1973 "Urban autonomy in a nomadic environment". *Journal of Near Eastern Studies* 32:201-15.
1977 Dimorphic structure and the parasocial element. Journal of Near Eastern Studies 36:181-98.

Saidel, B.A.
2000 "Matchlocks, Flintlocks and Saltpetre: the chronological implications for the use of matchlock muskets among Ottoman-period bedouin in the southern Levant". *International Journal of Historical Archaeology* 4/3:191-216.

Salzman, Ph.C.
1980 "Processes of sedentarization as adaptation and response" in *When Nomads Settle, ed. Ph. C. Salzman.* New York: Bergin. pp. 1-19.
2004 *Pastoralists. Equality, Hierarchy, and the State.* Boulder, CO. Westview Press.

Schumacher, G.
1888 *The Jaulan.* London: Bentley.
1886 *Across the Jordan: being an exploration and survey of part of Hauran and Jaulan.* London: Watt.

Singer-Avitz, L.
1999 *Beersheba – A gateway community in Southern Arabian Long-distance trade in the eighth century B.C.E.* Tel Aviv 26, 3-75.

Stanley, A.P.
1856 *Sinai and Palestine in connection with their history.* London: Murray.

Thesiger, W.
1959 *Arabian Sands.* London: Longmans, Green.

NABI MUSA: A COMMON *WELI* BETWEEN BEDOUIN AND FELLAHIN

Aref Abu-Rabia

Introduction

In the Negev area there is one legend that is particularly popular and with some variation in details is often repeated:

"According to the bedouin tradition, one day God wanted to take Nabi Musa's soul. Musa ran into the desert, and on the way he met a few bedouin digging a grave. He asked them who they were digging the grave for. They answered that their sheikh had died. Musa helped them dig the grave. When they finished digging, Musa asked them how they knew the dimensions of the sheikh. The bedouin answered that the sheikh's stature was similar to his, and asked Musa if he would lay down in the grave, so that they could make the dimensions more accurate. He agreed. The bedouin asked him to lie inside, to close his legs, and to keep his hands close to his body. As Musa finished doing as they asked him, the voice of God was heard from the heavens: 'Musa, Musa where are you? You cannot escape from me! I have sent you angels in the form of bedouin'. Musa begged God not to bury him in such a deserted and forsaken place. God answered: 'Next to your grave there will be wells of water. People will visit your grave, bring sacrifices with them, and light fires from the stones by your grave. Your grave will be a sacred place'. Then God took Musa's soul. The bedouin wanted to please Musa, and so they introduced the custom of visiting his grave in groups and even individually, one-by-one. They would make vows at the site, and keep them".

The bedouin, who maintain ritual visits (*zwara*) to Nabi Musa's tomb, believe that God blessed the place where Nabi Musa was buried with 'fire rocks' and water wells (*nareh men ihjareh wa-mayteh men ibyareh*).

The Nabi Musa tradition can provide insights into the sociology of the saint cult among bedouin and fellahin that can enrich our understanding of such cults elsewhere, and can throw light on their function and importance among these societies.

Review of the Literature

The cult of saints dates back to ancient times and can be found in many parts of the world, and assume many shapes. The most famous sanctuaries in pre-Islamic Arabia were those of Hubal,[1] al-Lāt,[2] Manāt[3] and al-'Uzza.[4] They had important spheres of influence. For example, the sanctuary of the goddess Manāt was located in Hudhail territory, but was frequented by al-Aws and al-Khazraj tribesmen and by others. The sanctuary at Mecca was visited not only by Quraish but also attracted pilgrims from other tribes. Consequently, many ties were forged between tribes from different regions, which facilitated transit between territories, mutual access to available pasture as well as trade, and meeting between different tribes.

Saint cults are prevalent in many parts of Islam (Levy 1962: 258). They are practiced in cities, villages, and

---

[1] Hubal: the name of an idol, which was worshipped at Mecca in the Ka'ba. One tradition told that Hubal was an idol of the Banu Kināna, worshipped also by Quraish, and had been placed in the Ka'ba by Khuzaima b. Mudrika. It is further related that the idol of Hubal was of red carnelian in the form of a man. The Quraish replaced the right hand which was broken, by a golden one. It was a custom to consult the idol by divination with arrows. After the conquest of Mecca by the Muslims, Hubal was removed from the Ka'ba and destroyed (Shorter Encyclopedia of Islam. 1974: 140).

[2] al-Lāt: this name refers to a specific female sun deity. She was worshipped by the various Bedouin, such as the Hawāzin. They were accustomed to take oaths by this idol. She had her principal sanctuary in the valley of Wajj near Ṭā'if, where the Mu'attib b. Ka'b were her priests and a white stone festooned with all kinds of decorations was her symbol. Among the Quraish, she was held in very high esteem, however, after the conquest of Mecca, al-Lāt was destroyed (ibid: 287).

[3] Manāt: an idol Arabian goddess. In Arabic her name corresponded to *maniya*- the allotted fate, specifically death. She was therefore a goddess of fate, especially of death. Her main sanctuary was a black stone among the Hudhailis in Qudaid, not far from Mecca on the road to Madina near a hill called Mushallal. She was, however, venerated by many Arab tribes, primary by the Aws and Khazraj in Yathrib. In Mecca she was very popular along with the goddesses al-Lāt and al-'Uzza; the three were regarded as Allah's daughters. Manāt was the oldest deity in Arabia, several tribes including the Aws and Khazraj assumed the *ihram* [a sacred state] at the sanctuary of Manāt and on the conclusion of the rites cut their hair there and dropped the *ihram*. Manāt was destroyed after the capture of Mecca (ibid: 325).

[4] al-'Uzza: an old Arabian goddess, whose name means 'the strong' or 'the powerful'. She was associated with the Ghaṭafān but her principal sanctuary was in the valley of Nakhla on the road from Ṭā'if to Mecca to which Ĥassan b. Thābit refers. It consisted of three acacia trees (*samura*) within one of which the goddess revealed herself. It also included the sacred stone and the so-called Ghabghab, a cave into which the blood of sacrificed animals was poured. Her cult spread among a number of Bedouin tribes: the Khuzā'a, Ghanm, Kināna, Bali, Thaqīf and Quraish. Outside Arabia al-'Uzza was worshipped especially by the Lakhmids of Ĥīra. Mundhir IV used to swear by her. A Lakhmid prince Nu'mān sent men to her so that she might settle a dispute. Her worship here assumed a particularly cruel character. Mundhir IV sacrificed 400 captured nuns in her honor, and on another occasion Mundhir IV sacrificed a son of the Jafnid Ĥārith, whom he had taken prisoner. Al-'Uzza was destroyed after the capture of Mecca (ibid: 617).

The shrine of Nebi Musa (photo by the author)

even tied to specific mosques (Canaan 1927: 2; Eickelman 1976: 112). Islamic scholars differ in their assessments regarding saint cults. Some scholars approve wholeheartedly of the practice, as in Linant de Bellefonds (1974: 355) who argues that 'the permissibility of visiting tombs was admitted very early on by *ijma'*, and is anchored in consensual canonical law practices. Scholars who oppose saint cults base their objection on the view that Islam is a strictly monotheistic religion, and they view saint cults as a relic or leftover from pagan culture, as a deviation from orthodox religion or as the product of a local custom.

Muslim hagiologists summarized the miracles of the saints in 20 categories. At the apex of the miracles that saints are purported to perform, scholars usually cite the power to resuscitate the dead (Goldziher 1971: 268-270). Others are: the curing of illnesses, sacred prayers that are always answered, divinations, performance of supernatural activities such as returning a lost object to its owner, seeing far away, and knowing what is going on somewhere else, dressing and undressing different objects. Grunebaum (1951: 67), following Goldziher, views the cult of saints as a striking deviation from the genuine prophetic tradition. Lazarus-Yafeh (1967: 332) also agrees with Goldziher, but notes that the custom of visiting the saints' tombs began in ancient times. The cult of marking and visiting saints' tombs is common in all the countries of the Middle East and North Africa. Many of the saints' tombs are common to the Mizrahi Jews and Arab Muslims (Ben-Ami 1984; Bilu 1993: 53-60; Geertz 1968: 23-55). There have been Muslim scholars who fought the custom of saint cult, but all their attempts have failed. Gellner (1972: 59) believes that the saint cult is an alternative to official Islam in regions where the present form of Islam takes a minimalist form.

Most Muslim scholars claim that the worship of saints contradicts the tenets of Islam, while underscoring that a visit to a gravesite or cemetery is not counter to Islam – provided that there is no weeping, keening, scratching of faces and so forth. The purpose of the visit is to remind the visitors that there is no escape from death.[5] It is forbidden to worship the saints; the performance of ceremonies, such as the lighting of candles, by their graves, as well as the making of special requests for the cure of diseases, are likewise forbidden. All who need help must direct their prayers solely to Allah, because He alone is omnipotent (Shaltut 1966; Sharabasy 1970). The tombs of saints can be visited in order to partake in their sanctity, to learn a lesson and to accept the inevitability of

[5] "Where so ever ye may be, death will overtake you, even though ye were in lofty towers" (Quran 4:78).

death. At first, the prophet Muhammad forbade visits to the tombs of saints, but later permitted them, provided that they should be carried out in the spirit of Islam. The prophet recommended visits to graves and particularly to those of one's parents on Fridays. He spoke of this as an important good deed; on this occasion, it was suitable to read verses from the Quran and particularly from the Opening Sura, the Unity, the Daybreak, and the Mankind (Al-Ghazali 1992: 484-493). There is, however, a tremendous gap between this concept of early Islam and the kind of veneration of saints by believers which involves the actual practice of the saint cult, and which views the dead as mediators between themselves and Allah, in order to satisfy or realize the supplicant's needs or desires.

The saints fulfil an important function in the life of the Arab Muslims in the Middle East and North Africa. With the passage of time, many legends and narratives have sprung up around each tomb and its saint, including miraculous tales of the willingness of the saint to help the downtrodden, to cure illnesses, to reveal the future and so forth. Such venerated personalities are purported also to be able to find objects that have disappeared or been hidden. They are believed to employ their powers in a positive manner to heal the sick, solve thefts and counter witchcraft and sorcery (Al-Krenawi 2000; Ben-Ami 1984; Bilu 1993: 53-60; Talmon-Heller 1996: 95-109).

In Palestine, the local saint serves as a tangible figure for the Muslim, who embraces the saint's tomb when he is having trouble, or seeking a cure for an illness. Sometimes the sick sleep overnight at the site of the tomb, or even for a longer period. Women visit the tomb in order to cure infertility, and ask to give birth to male offspring. Handicapped persons with incurable medical problems visit the tomb of the saint to request a cure. Inhabitants living in the proximity of the tomb ask the saint to plead with Allah on their behalf for rainfall during drought years.

The majority of the visits to Nabi Musa's tomb are during the spring, which is the most pleasant season of the year (Canaan 1927). The encounters between bedouin and fellahin at the different ceremonies are a major source of cultural cross-fertilization.

Visiting the shrine of saints is an important event in the social life of the bedouin of South Sinai (Marx 1985: 106), those of Cyrenaica (Evans-Pritchard 1949: 66-67; Peters 1990: 16, 37-39,107-108), and Morocco (Gellner 1969, 1983: 114-130, 207-220). According to Marx (1985: 104-131) throughout the year individual bedouin visit tombs whose resident saint is believed to have the power to help them with their problems. All major shrines are classified by the bedouin as *maqam*, literally 'place', a term widely used in the Islamic world to denote an important holy place, mostly, but not always, associated with a saint's tomb. Most shrines are surmounted by white domed buildings, crowned with a crescent, the emblem of Islam. Usually there are trees and a well or cisterns near the shrine.

Among the bedouin of South Sinai people go on pilgrimage either to redeem vows made when in danger, or to make requests. As the saint is only mediator for their requests to the almighty God, a visit to any holy tomb will do (Bailey 1982: 65-88; Ben-David 1981: 107-118; Levi 1987: 385-424; Marx 1985: 117; Shuqayr 1916: 352-397).

There is hardly a territory that does not have a holy tomb that fulfils most of the bedouin's religious needs. This is one reason for the abundance of holy sites in Palestine. According to Marx (1985: 123) the use of a saint's tomb as a symbol of tribal ownership indicates that territorial rights are protected by supernatural sanctions. Bedouin store property, such as tents and farming implements inside or near a holy tomb, secure in the belief that such property is protected by the saint. Any goods deposited there will never be stolen (Palmer 1871: 119).

The saint represents both the tribe as conceptualized by the bedouin, and the wide external world. His function is to mediate between the two worlds which together sustain the bedouin (Marx 1985: 1126-127). The bedouin make their pilgrimages to strengthen three levels of ties: those between the members of one tribe, those between the tribes of southern Sinai and those between the tribal members and all of Islam (Marx 1977: 14-22).

The tombs of saints are important both to bedouin and to their animals. Every visit to the tomb must be consecrated by the sacrifice of a sheep or goat, sealing the bond between the bedouin and his flock. Since the flock constitutes the primary link between the bedouin and the saints, and since every sacrificial ceremony is followed by a festive family gathering, the flock contributes to the cohesion and consolidation of the family in different forms and ways. The flock is the connecting link not only between human beings, but also between individuals and their God, their prophets, their saints, and venerated pious figures (Abu-Rabia 1994: 90-91).

The saint's tomb is covered by a shroud made of white or green cloth, on which sometimes the Islamic creed and the saint's name are embroidered. There are usually several layers of shrouds, for each time a shroud is worn and torn it remains in place and a new one is placed on top of it. Bedouin used to take the shroud remains for treatment of illnesses and for protection against jinns. (The spirit of the deceased is believed to be carried on through his clothes. They are believed to protect against the evil eye and other bad omens. Bedouin in Sinai, after burying their dead leave the deceased's clothes on the grave for passers-by to take. The bedouin near al-Tur in the Sinai leave the clothes on a tree or rock near the cemetery (Shuqayr 1916: 395-396). According to Bailey (1991: 65) it is considered meritorious to leave the clothes of a deceased bedouin (who would be buried in a shroud) on his grave, anticipating that a poor person might find them).[6]

The bedouin are accustomed to swear by the tombs of the saints and also swear on them; and they believe that whoever takes a false oath on one of the holy tombs will die or suffer some great tragedy within a short while(Ashkenazi 1957: 88-102). The saint's tomb is also called the tomb of the *weli, weli* being the bedouin word for a saint or holy person.

The custom of visiting the tombs of saints still exists among the bedouin of the Negev and of Sinai (Al-Aref 1933: 252-264; Levi 1987; Haj Hamd Abu-Taha, *pers. comm.*, 9 February 1984; Haj Hammad Abu-Rabia, *pers. comm.*, 14 May 2000). Generally, it is the family who visits the tomb of a particular saint or ancestor, either to fulfil a vow made in his memory, or in times of illness. The supplications and prayers are directed either to the saint or to Allah through the mediation of the saint. In the evening, the family and relatives gather together in the tent and slaughter a sheep or goat to guarantee that they will fulfil the vow. Sometimes, the slaughter is carried out at the graveside (Abu-Rabia 1983: 1-15; 1999: 21-32, 82-93). It is a unique opportunity for the bedouin to go through a rite of passage during the saint's pilgrimage, beginning in a familiar place – the tribe's regular habitat, going to a far place – the saint's shrine; then, returning to a familiar place – the tribe.

Nabi Musa's Shrine

Nabi Musa's shrine is situated seven km southwest of Jericho, just south of the road leading from Jerusalem to Jericho (Canaan 1927: 193-194; Conder & Kitchner 1883: 220; Sahhab 1990: 638). The saint is highly honoured by all Muslims of Palestine. People come to his feast from villages and towns from all over Palestine.

The traditional burial place of Nabi Musa, by Jericho, coincides with the general traditional view in Palestine. When Muhammad made his midnight journey through the sky from Mecca to Jerusalem and ascended to the heavens for a divine meeting with Allah,[7] on his way back he saw Nabi Musa praying by the reddish hillock – *al-Kathib al-Ahmar*. Traditions that were attributed to Nabi Musa's tomb include claims such as: shadows and lights appearing at the site as a sign of its holiness, and birds that sensed the sacred aura of the site and picked the leftovers from the ground (Sadan 1993: 231-245).

According to Canaan (1927: 238-239), most saints are supposed to wear the clothes of their locality. It is interesting to note three colours red, green and white. White predominates, while green is the holy colour of Islam. In the case of Nabi Musa, many have sworn that they have seen him with a greenish halo surrounding his face.[8]

It should be noted that the aftermath of the Crusades sparked a religious awakening amongst the Muslims, who started building and reconstructing many religious structures in Palestine.[9] The fear of a renewed Crusade drove Muslim political leaders to concentrate large armed forces in central cities and strategic sites – such as Jerusalem, Ramleh, Gaza, and Acre – at dangerous periods of the year. This objective was partially accompanied by the foundation of centres that were sacred to the Muslims, such as known saints' tombs that were visited by the Muslims on set holidays. The spiritual leaders supported the rulers, since they saw their actions as a vehicle for spiritual awakening. The visitors were usually armed and mounted on horses. The *Waqf* [10] dedicated much land to the important tombs, and took care of the expenses of the visitors and the cost of festivities. In addition, the roads leading to the tombs were repaired, and *khan*s were built near them to ease access. In this fashion the Nabi Musa tomb was also built. The structure was first erected by the Royal Sultan Baybars I,[11] the fourth royal sultan. He also stood at the head of the festivities. After his pilgrimage to Mecca, Baybars paid visits to Damascus, Jerusalem, and Hebron. There, in 1270, he built the tomb's structure, and by it a mosque. The site was renovated and expanded in 1480. Later the site was surrounded by a wall and other structures were added. The *maqam* received its present shape only in the 19th century, when the *khan* was renovated and enlarged in order to receive all the people who came for the pilgrimage and festivities. The complex consists of several structures, including the holy shrine which is surrounded by rooms, together forming a wide yard. The room of the tomb itself is bordered on the north and the east sides by an arched yard. Many articles of clothing[12] are tied to bars on the windows, and over the door is a sign testifying that the site was renovated by 'Abdallah Pasha, who rebuilt the place in 1867' (Sadan 1993: 231-245; Tamari 1979: 153-180). The structure was intended to commemorate the presence of the Muslims in Palestine.

---

[6] "A poem says "On your friend they have heaped the turf, and over his gravestone flung his cloak".

[7] "Glorified be He Who carried His servant by night from the Inviolable Place of Worship [Mecca] to the far distant place of worship [Jerusalem]…." Quran 17:1.

[8] Compare a number of legends that are told on the authority of al-Nawār bint Mālik, such as a vision that celestial light surrounded the urn in which 'Ali [bin Abi Ṭālib]'s head was kept and that a white bird fluttered around it (Al-Ṭabari, II, p.369). The famous saint Rābi'a al-'Adawiyya (717-801)wa envelopged by a radiance while she prayed (Hitti 1951:439; Smith 1928).

[9] The fall of Jerusalem by the Crusades was in 1099 (Hitti 1951:480).

[10] *Waqf*, pl. *awqaf*, an Islamic religious endowment. Commonly a pious foundation, which is defined in various ways in the *Shari'a*. An object which while retaining its substance yields a usufruct and of which the owner has surrendered his power of disposal with the stipulation that the yield be used for allowable good purposes.

[11] Fourth sultan of the Mamluk period (1250-1517), al-Malik al-Zāhir Baybars (1260-1277).

[12] To be used for medical purposes, after they have absorbed the healing powers of the shrine.

A common decoration on the structure of the tomb is an upraised open palm – a *khamsa*. According to common Muslim belief, it is the hand of Fatima, the daughter of the Prophet Muhammad, and it is believed to have special spiritual powers.[13]

Nabi Musa's Zwara

The *zwara*, the visiting of the saints' tomb on special occasions, holds a special and unique importance. The most common visits are during the Nabi Musa celebrations in spring. Visitors usually come in groups. Celebrants arrive at Nabi Musa's tomb riding donkeys, mules, camels, horses, and even on foot. At times, the visit involves a lengthy journey, and therefore the hardships of the trip become an integral part of the pilgrimage experience, and a form of purification in expectation of meeting the saint.

Since the *zwara* to Nabi Musa typically lasts several days, the pilgrims equip themselves with supplies of food, dishes, water, tents, blankets, clothes, and lambs for slaughtering. Visitors habitually conduct general and private prayers by the tomb of Nabi Musa, touch and kiss it, and light incense and lamps filled with olive oil. They put olive oil, incense, food, candles, and even water by the tomb. They also hang a white cloth on the tomb, or on a nearby tree. This cloth serves as a constant reminder to the saint not to forget the visitors' supplications. In this manner, a tie between the saint and the visitors is forged. The clothes also absorb the supernatural powers of the saint,[14] and pass them on to the owner of the cloth. Water is employed as a medium for blessing individuals that suffer from a particular disease. The visitors bring objects with them that they place near the tomb for those who couldn't make the *zwara* to the tomb. A patient suffering from the evil eye, evil spirits, jinn, or madness is treated by fumigation with incense taken from Nabi Musa. Bedouin women used to take a small stones from Nabi Musa and hang them as a necklace (*qladih*) around their neck and down the abdomen over their womb in order to hasten a difficult labour.

A narrative, retelling what befalls on an individual *zwara* is enlightening. The author was given the following details by Abu-Mhammad – a bedouin from the Negev, who together with his wife and five children made a pilgrimage to Nabi Musa:[15] Abu-Mhammad's wife had made a vow to the honour of Nabi Musa when one of her children – Suilim – was seriously ill. One Friday morning, Abu-Mhammad, his wife and children set out in their pickup truck to visit Nabi Musa's tomb. Abu-Mhammad's wife tied a white scarf (*rayih baydha*) to the front of the vehicle in a prominent place so that during the journey it could wave in the wind and be seen from afar. After driving for two hours they arrived at Nabi Musa's shrine near Jericho, and got out of the vehicle. They approached the tomb and read *Surat al-Fatiha*. At the end of the reading Abu-Mhammad's wife hung a piece of white cloth near the tomb, she then produced some incense (*bakhour*) that she had brought with her, and lit it in the area of the courtyard of the shrine. She took out incense, olive oil, rings and coins and laid them out on the ground. She uttered personal prayers for health, happiness and wealth for members of her family, and then in closing, lowering her voice, added a personal supplication for herself. Afterwards, she held her children and together with them circled the tomb, touching and even kissing it. With one hand she held Suilim, and with the other hand she touched the tomb and said: "Oh, ya Nabi Musa, O ya *kalim* Allah,[16] O our protector, this day I place my children under the protection of God, and under thy protection. I put my son Suilim under the protection of God and thy protection. I pray thee, guard them from the evil eye, jinns, spirits, and from sickness that are sent to afflict people by devils on the face of the earth and under it, and guard them also against witchcraft (*sihr*) and sorcery (*'amal*).[17] Oh Nabi Musa, *jinak la tkhayib amalna fiyk* – we invoke you, please don't disappoint us, today we have come to visit you and to fulfil the vow that we have sworn. I pray thee, do not forget us, who are thy children under thy protection and who are thy servants. Remember us always, as a devoted father never forgets his children. Oh, Nabi Musa, give us good health, much happiness and wealth, long and good lives. Bless our livestock and our camels, bless (*barik*) our *zari'* (young standing green crop), bless our wells and cisterns, bless our children and our descendants, our fathers and mothers, our grandfathers and our grandmothers. Oh, Nabi Musa, have mercy on us, and protect us from all bad doers and evils. O Nabi Musa please, bless me to be pregnant with a new son".[18] At the end of these invocations and supplications she read again *Surat al-Fatiha*. While she practiced her ritual, her husband Abu-Mhammad took the Quran near the tomb

---

[13] According to the Christian belief, it is the hand of the Virgin Mary, mother of Jesus. According to the Jewish belief, it is the hand of divine supervision.

[14] Compare to the relics (*athār*) of the Prophet Muhammad, considered to be some sort of amulets that are kept for curative purposes and blessings (*littabarruk*) (Goldhizer 1971:324). In Mecca the miraculous staff of Moses was kept as a sacred relic, as well as the horn of the ram which served Ibrahim as a vicarious sacrifice. Both these sacred treasures were covered with gold and mounted with precious stones. The Qarmatians plundered these treasures in the sack of the city in 317 Hijra and they have not been seen since (Amir Khani 1993:95-121).

[15] Personal communication from Abu-Mhammad, 28 May, 1998.

[16] Interlocutor of God.

[17] From the evil of women blowers upon knots, a common form of witchcraft in Arabia whereby women tie knots in a cord and blow upon them with an imprecation. See Surat al-Falaq – The Daybreak, Quran 113:1-5.

[18] The wicks burnt at Nabi Musa, and swallowed by sterile women were believed to cure their condition. Others carry these wicks as an amulet (*hijab*) against sterility. Women used to cut off a piece of the tomb-covering of Nabi Musa and make it into a cup for a sickly child or for a child whose mother has lost most of her male children. A thread which has been passed around the tomb of Nabi Musa (and which has thus the length of its circumference) worn around the waist of a woman suffering from continuous miscarriages, prevents such an evil in the future (Canaan 1927: 113). In Bedouin society, to be pregnant is to be blessed, and by contrast, to be sterile and is to be cursed.

and read some verses. Then he prayed three *raka't.*[19] When the family had finished their *zwara*, they drove back to the tribe. There, in the evening, they slaughtered three lambs for the invited relatives and guests. Men cooked the meat, while women prepared bread (*khubz*) and rice with saffron (*hawayij*). Men sat in the men's section, and the women in the women's section. The male children were roaming about and playing, but they were also taking notice of the ceremony of the sacrifice. Some of them joined the men, others the women. Girls joined only their mothers, while the boys had both options. The boys listened to both the women's and the men's conversations, that related legends and miracles about the attributes and deeds of Nabi Musa.

When the food was ready, bread was put in the bowls and soup (*maraqa*) poured on it, rice added, and pieces of meat, then more soup. The bowls were passed round, to serve women and men at the same time. At the end of the meal, in accordance with Abu-Mhammad's request, all those partaking of the meal ritual held their hands out in front of them with the palms upwards, and recited *Surat al-Fatiha*. Everyone responded "*Amin, zwaritku magbula*", which means 'your visit to Nabi Musa has been accepted'. A similar ceremony took place in the women's section. Then Abu-Mhammad, with the help of his brothers, returned the bowls to the kitchen. The older children of Abu-Mhammad and his brothers had prepared water, soap and towels, taking turns to pour water over the hands of those who had eaten. (It should be noted that before eating the food, they washed their hands with water without soap.) After having drunk coffee and tea, the guests left for their homes.

All this makes clear that the ceremonies of visiting saints' tombs have established a relationship of psychological-therapeutic dependence on the part of the bedouin with regard to saints. This dependence is deeply rooted in their psyche and reinforced and legimitized by the bedouin culture.

Since the bedouin move with their livestock and camels from one site to another, they do not have much opportunity to visit the Nabi Musa shrine. Their visits to the shrine are usually limited to the spring, when they are pasturing in the neighbourhood, and which is also the time of the annual celebration. Sometimes they pay other visits during the year, but this is not common. Instead, they visit other holy tombs, which are found in the whole of Palestine. Since the fellahin have less opportunity, and less urge to visit the shrines in the desert that are frequented by the bedouin, the spring celebration is the time when the bedouin and farmers generally meet and celebrate together at the Nabi Musa shrine. This is also the time when certain rituals, shared by both bedouin and fellahin, are conducted

[19] A bending of the torso from an upright position followed by prostrations in Muslim prayer ritual.

Circumcision Rituals

Circumcision is a very common ritual at Nabi Musa's shrine. The child is dressed in cloths, a blue bead, amulet, and alum are placed on him, to protect him from the evil eye and spirits (*nafs*). Usually the barber who is also the professional circumcision surgeon performs the small operation with the utmost skill. Women dance and sing: *ya shalabi khiff idak almus la igta' idak*: O circumciser, make your hand light in order that the knife will not cut your hand.

The wound is dressed with dried plants,[20] powdered and sometimes mixed with olive oil taken from the lamps of the *maqam*. Many believe that the latter alone is the best cure. Sometimes the barber uses a drying antiseptic powder.

Haj Salamih[21] told me that the bedouin used to sacrifice one sheep/goat for each circumcised boy. But when they know that a lot of guests will attend the ceremony, they often slaughter more than one animal. In large ceremonies of notables, the number of sacrificed animals can be up to twenty. The sacrificed animal devoted to the circumcised boy is called *'aqira.*[22] In the past, the heads of the sacrificed animals were collected and placed near the men's tent at a distance to be used as shooting targets *(ṭu'm lil-bārud*). The man, usually from among the guests, who succeeded in hitting the target, 'won' a large portion of meat and at the same time the praise from the women's *zagharut*: the uttering of shrill shouts of joy. This used to be the 'national bedouin sport' but today celebrants don't dare to fire weapons in ceremony, fearing that the police will accuse them of 'causing danger to the public', and confiscate their rifles. 'I really don't know how he could endanger the public, while this is exactly what the public want to do and hear', muttered Haj Salamih.

Since the bedouin have less opportunity to meet professional circumcisers than the fellahin, who live together in villages, the Nabi Musa celebrations are the ideal occasion to bring their male children for circumcision, because here they find skilled circumcisers. The circumcisers are generally from the settled community, and so this is another occasion where both communities meet and interact. The sacredness of the place also is supposed to help the circumcision wound heal faster and better.

[20] A powder (*rushush*) made from the dried leaves of plants is placed on the penis at the time of circumcision to reduce bleeding, heal the cut and reduce the danger of infection. Some of these plants are: Artemisia judaica L. (*b'aythran*), Artemisia herba-alba (*shih*), Achillea fragrantissima (*qisum*), Retama raetam (*ratam*), Coridothymus reichb (*zahaify*), Cleome droserifolia (*samwih*), Ballota undulate (*ghassih*), and Cinnamomum zeylanicum (*qirfih*).

[21] Personal communication March 10, 1990.

[22] *'Aqara*, to cut the legs or hock-tendon (*'arqub*, pl. *'arāqīb*) of sheep, goat or camel with a sword while it was standing, to fell the animal in order to cut its throat. *'Aqira* is the animal slaughtered for circumcision. It is a symbolic term to signify cutting the boy's foreskin by a sharp knife/straight razor.

Rain Rituals

Sometimes, in years of drought, people who live in villages close to the shrine can implore Nabi Musa's intercession to produce rain. This involves a special ritual.

The rain ritual begins when the people gather in a procession that marches around the village. When the gathering reaches Nabi Musa's tomb, they encircle the site, then enter the *maqam* and recite the *al-Fatiha*. The participants raise their hands to heaven and pray for rain. A large doll in the image of *Um al-ghayth*, the mother of rain, is carried throughout this procession.[23] The doll is made of two pieces of wood that are fastened together in the form of a cross as a manikin, and dressed in female attire. In this procession the women carry one or more grain sieves (*ghurbal*, pl. *gharabil*), flour sieves (*munkhul*, pl. *manakhil*), and a hand mill (*tahuneh*) all the way. These are the most important utensils for preparing bread. All are carried empty to show how poor their owners are, since they do not even possess the most vital necessities of life. During these processions the peasants sing songs, one of which is:[24]

Ya rabbi lish hal-kanneh wakalna 'uruq al-kirsanneh [25]
O Lord, why this calmness while we have eaten the roots of eryngo.
Ya rabbi lish u lish wakalna 'uruq al-khurfish [26]
O Lord, why, why – and we have eaten the roots of thistles!
Ya rabbi lish hal-qa'deh wakalna 'uruq al-ja'deh [27]
O Lord, why this neglect, while we have eaten the roots of the mountain germander".

Around Nabi Musa there are large cisterns in which rainwater is collected for use in the feast days. Bedouin and fellahin believe that these cisterns are often filled in a miraculous way. When the rainfall during the winter is scanty and the cisterns are not full of water, it is said that Nabi Musa sends a heavy downpour of rain, which falls only over his sacred area, shortly before the beginning of the feast.

The ceremony asking for rainfall exists in the Arab culture throughout the Middle East and North Africa (Abu-Zahra1988: 507-529; Van Binsbergen 1985: 209-210). There is even a special prayer (*Salat al-Istisqa'*)[28] requesting rainfall in the Muslim religion. When the Medinians faced starvation because of a drought, 'Ā'isha advised them to make an opening ceremony towards the sky on the grave of the Prophet. Thus the sacred grave was brought into direct contact with the angry heavens. Abundant rain followed immediately causing the grass to shoot up and the cattle to thrive. It was said that in the 5th century, when severe drought prevailed in Samarqand, the community made a pilgrimage to Khartank in order to pray at al-Bukhari's tomb, and the rain appeared in downpour. In Morocco, in the year 711, when the country suffered from drought, prince Abu-Sa'id made a pilgrimage with his community to the tomb of the saint Abu Ya'qub al-Ashqar. God sent a downpour and revived the dry lands. One time in Egypt the Nile failed to rise and irrigate the dry land, and the people were facing drought and famine; Nafisa[29] handed her veil to the despondent citizens so that they might throw it into the Nile. Immediately, the level of the river began to rise and the people witnessed an abundance as rarely seen before (Goldziher 1971: 277-278).

In the rain rituals the bedouin pray to God to bring plenty of rain in order to have more water for them and their animals, as well as pasture for their animals. Most of the fellahin need this rain for themselves and their small gardens (*bustans*). Bedouin also cultivate their lands, but in case of need, they buy barley and wheat from the fellahin. The fellahin, in their turn, buy from the bedouin milk products such as dried cheese (*'afig*), and *samna* (processed butter). Here again is an occasion where the bedouin and the fellahin share a common need that can be fulfilled by appealing to the common Saint.

The two groups share a world in which they are dependent upon each other for their subsistence, in which both need the same rain. Therefore, however different their lifestyles may be, they need each other, and this mutual dependence becomes clear at the shrine of Nabi Musa.

Appendix: Nabi Musa' Rocks

The bedouin name for bitumen/asphalt is *hummarah*. Bitumen – a form of congealed petroleum – was collected by the bedouin at least once a year in the form of blocks that can be found floating on the Dead Sea. The petroleum oozes into the Dead Sea via springs or ground water, and then congeals in the salt water into bright

[23]This custom is more common among the bedouin than among the peasants, especially among the bedouin of Jordan. The doll can be compared to *Mother Tambu* inTunisian rain rites, Abu-Zahra 1988: 522. See also Musil 1928,10-13, where he describes the procession and cites the songs that are sung.

[24] Cited from Canaan. The scientific names given by the author.

[25] *Kirsanneh/ kursanneh*: Eryngium creticum Lam; or Vicia ervilia.

[26] *Khurfish al-Hamir*: Notobasis syriaca (L.) Cass.; or Cynara syriaca.

[27] *Ja'adeh*: Teucrium polium L.

[28] Salat al-Istisqa': The special ritual to be observed in the prayer for rain consists of: 1. a prayer of two *rak'a* must be performed in the morning outside the town. 2. the faithful should put on ordinary dress, without elaboration or luxury. 3. the prayer is followed by two public addresses (*khutba*), of which the first is accompanied by a turning of the cloak (a sympathetic rite to produce a change in the weather); 4. the *du'a'* which follows the prayer is a supplication for rain; 5. the usual *takbir* is replaced by an invocation intended to improve God's pardon (*istighfar*). This prayer should be completed by a series of pious actions recommended to the faithful (fasting, almsgiving) (Alfred Bel, in *First Encyclopedia of Islam*, vol. 3.1993:562-563).

[29] Al-Saiyida Nafisa (762-824) daughter of al-Hasan b. Zaid b. al-Hasan b. 'Ali b. Abi Talib; and daughter-in-law of the Imam Ja'far al-Sādiq. She was famous for her piety and zeal in religious practices. She moved from Arabia to Egypt. Her miracles are without number. The people of Cairo regarded the tomb of Nafisa as a privileged place where prayers are said with certainty of fulfillment. The legends of Nafisa represent a special type of legend about women saints which extend to both the east and west of the Islamic world (Goldziher 1971: 277-278).

reddish-black blocks with a pleasant smell. An easterly wind would blow the asphalt blocks towards the western shore. The bedouin would break the large blocks into smaller blocks that could be loaded onto camels. They sold these blocks to Arab traders in Hebron, Jerusalem, and elsewhere (Hajj Hammad Salem Abu-Rabi'a, *pers. comm.*, 21 August, 1998). In Palestine, very small quantities of bitumen were thus collected, a fact reflected in its high price: in 1872, a camel-load of bitumen fetched 500 grush in Jerusalem and Bethlehem. Such trade was an important source of income for the bedouin tribes that controlled the western shore of the Dead Sea.

The bedouin made tar pastes out of the bitumen to spread on tree and vine stumps after pruning, and to mend cracks in pottery. It could be boiled in olive oil over a flame and chewed like gum. Medicines were also made out of the blocks for dental therapy, and it could be used to hold the jaws and to treat inflammation of the gums. Bitumen was also used to treat children's stuttering, by 'freeing their tongue' so they could speak more easily, to treat fever, diarrhoea, and inflammation of the uterus. It was also used to treat skin diseases in animals and humans (Abu-Rabia 2001: 56-57). The author has the impression that the bedouin generally had more faith in the magic powers of Nabi Musa's rocks than the fellahin.

In contrast to the paltry bitumen production, larger quantities of oil were produced from 'Moses rocks' (*ihjar Musa*), the oil-bearing bituminous shale rocks found in the vicinity of the Nabi Musa shrine.

Before the twentieth century, bituminous shale was only rarely used to produce heating fuel, and then usually by bedouin who chanced to be in the vicinity of such rocks. The extraction of fuel from shale on a regular basis began only during the great fuel shortage of World War I. The Germans produced aviation fuel from bituminous rocks at Makarin, east of el-Hamma, along the Hejaz railway. The Ottomans used petroleum from bituminous shale to fuel locomotives, and to power their steam boat on the Dead Sea (Avitsur 1976: 274-77; Bar-Zvi, Abu-Rabia, and Kressel 1998: 50).

According to Canaan (1927: 88, 110), stones around Nabi Musa are black and contain some bitumen, so they burn when put in a fire. They show their special power in being able to burn. They are black because they contain a bituminous substance. The black stones are cut in square or triangular forms, a protective talisman is inscribed, and they are carried as a protective amulet (*Hijab*, *Hirz*).

In Palestine, bitumen was used to seal cracks and make medicines; while another possible use was in the making of dye and paint. Eventually the demand increased, particularly after the British conquered Palestine in 1917. This gave rise to the need for the production of much larger quantities, and from cheaper sources. These needs were met by bitumen quarried in the mines of Hatzbaya, Lebanon, from which the material was also exported overseas (Avitsur 1976: 274-77).

It should be noted that natural bitumen was widely used in the ancient world, as far back as about 9000 B.C. In ancient Mesopotamia it was used as building material such as mortar and cement, as paving for roads, to waterproof pipes, as insecticides, for medicinal purposes and magic rituals as well as for making fire. Dead Sea bitumen was an ingredient in balms in Egyptian mummies (Connan, Nissenbaum, and Dessort 1992: 2,743-59; Nissenbaum 1978: 837-44). According to Brewer and Teeter (1999: 5) "In the seventeenth century eating dried flesh of a mummified corpse was considered a medical cure-all; the flesh and wrapping of Egyptians mummies were covered with a dark resinous substance, likened to bitumen, and were therefore referred to as mummy".

Bibliography

Abu-Rabia, A.

1983 *Folk Medicine among the Bedouin Tribes in the Negev*. Sde Boqer Campus, Ben-Gurion University of the Negev.

1994 *The Negev Bedouin and Livestock Rearing: Social, Economic and Political Aspects.* Oxford: Berg.

1999 *Traditional Bedouin Medicine*. Tel Aviv: Mod publishing (Hebrew).

2001 *Bedouin Century, Education and Development among the Negev Tribes in the Twentieth Century*. New York and Oxford: Berghahn Books.

Abu-Zahra, N.

1988 The Rain Rituals as Rites of Spiritual Passage. *International Journal of Middle East Studies* 20: 507-29.

Al-Ghazali, M.

1992 *Ihya 'ulum al-din*. Beirut: Dar al-Hadi.

Al-Krenawi, A.

2000 *Ethno-Psychiatry among the Bedouin-Arab of the Negev*. Tel Aviv: Hakibutz Hameuchad Publishing House (Hebrew).

Amir Khani, M.B.

1993 *Qarmatiyān-i Bahrayn va Faṭimiyān*. Tihran: Surush.

al- 'Aref, 'A.

1933 *Al-Qada bayna al-Badw*. Jerusalem: Bayt al-Maqdes.

Ashkenazi, T.

1957 *The Bedouin: Origin, Life and Customs.* Jerusalem: Mas.

Avitsur, S.

1976 *Daily Life in Eretz Israel in the Nineteenth Century*. Jerusalem: 'Am Hasefer.

Bailey, C.

1982 Bedouin Religious Practices in Sinai and the Negev. *Anthropos* 77: 65-88.

1991 *Bedouin Poetry from Sinai and the Negev.* Oxford: Clarendon.

Bar-Zvi, S.; Abu-Rabi'a, 'A.; and Kressel, G.

1998 *The Charm of Graves: Mourning Rituals and*

Tomb Worshipping Among the Negev Bedouin. Tel Aviv: Israel Ministry of Defence.

Bel, A.
1993 *First Encyclopedia of Islam*, vol. 3.

Ben-David, J.
1981 *Jabaliya, a Bedouin Tribe under a Monastery's Patronage*. Jerusalem: Kanna.

Ben-Ami, I.
1984 *Saint Veneration among the Jews in Morocco*. Jerusalem: Magness.

Brewer, D. and Teeter, E.
1999 *Egypt and the Egyptians*. Cambridge: Cambridge University.

Bilu, Y.
1993 *Without Bounds, the life and death of Rabbi Ya'acov Wazana*. Jerusalem: Magness.

Canaan, T.
1927 *Muhammedan Saints and Sancuaries in Palestine*. London: Luzac.

Conder, C. and Kitchener, H.
1883 *The Survey of Western Palestine, III*, London: Harrison & Sons.

Connan, J.; Nissenbaum, A.; and Dessort, D.
1992 Molecular archaeology: Export of Dead Sea asphalt to Canaan and Egypt in the Chalcolithic-Early Bronze Age (4th-3rd millennium B.C.). *Geochimica et Cosmochimica Acta* 56: 2743-59.

Eickelman, D.
1976 *Morocan Islam: Tradition and Society in a Pilgrimage Center*. Austin: University of Texas.

Evans-Pritchard, E.E.
1949 *The Sanusi of Cyrenaica*. Oxford: Clarendon.
1993 *First Encyclopedia of Islam*. Leiden: Brill.

Geertz, C.
1968 *Islam Observed, Religion Development in Morocco and Indonesia*. Chicago: University of Chicago.

Gellner, E.
1969 *Saints of the Atlas*. London: Weidenfeld & Nicholson.
1972 "Political and Religious Organization of the Berbers of the Central High Atlas" in *Arabs and Berbers; From Tribe to Nation in North Africa*, eds E. Gellner and C. Micaud. London: Duckworth. pp 59-66.
1983 *Muslim Society*. Cambridge: Cambridge University.

Gibb H.A.R., Kramers, J.H.
1974 *Shorter Encyclopedia of Islam*. Leiden: Brill.

Goldziher, I.
1971 "Veneration of Saints in Islam" in *Muslim Studies*, vol. II. London: George Allen & Unwin Ltd. pp. 255-341.

Grunebaum, G.E. von
1951 *Muhammedan Festivals*. New York: Schuman.

Hitti, P.
1951 *History of the Arabs*. London: MacMillan & Co.

Lazarus-Yafeh, H
1967 "Saints Worship" in *Studies in the History of the Arabs and Islam*. Tel-Aviv: Reshafim Publishing House. pp. 331-332

Levi, S.
1987 *The Bedouin in Sinai Desert*. Tel-Aviv: Schocken Publishing House.

Levy, R.
1962 *The Social Structure of Islam*. Cambridge: Cambridge University.

Linant de Bellefonds, Y.
1974 "Kabr" in *Encyclopedia of Islam*. New Edition. Leiden: Brill. pp. 352-355.

Marx, E.
1977 "Bedouin Pilgrimages to Holy Tombs in Southern Sinai". *Notes on the Bedouin*, VIII: 14-22.
1985 "Tribal Pilgrimages to Saints' Tombs in South Sinai" in *Islamic Dilemmas: Reformers, Nationalists and Industrialization,* ed. E. Gellner.. New York: Mouton Publishers. pp. 104-131.

Musil A.
1928 *The Manners and Customs of the Rwala bedouin*. New York: National Geographical Society

Nissenbaum, A.
1978 "Dead Sea Asphalt-Historical Aspects". *American Association of Petroleum Geologists* 62: 837-44.

Palmer, E.
1871 *The Desert of the Exodus, I*. Cambridge: Deighton, Bell and Co.

Peters, E.L.
1990 *The Bedouin of Cyrenaica*. Cambridge: Cambridge University.

Sadan, J.
1993 "A Legal Opinion of a Muslim Jurist Regarding the Sanctity of Jerusalem". *Israel Oriental Studies*. Vol. XIII: 231-45.

Sahhab, F.
1990 *Encyclopedia Palestina*, Vol. IV. Bayrut: Encyclopedia Palestina orporation.

Shaltut, M.
1966 *Al-Islam 'aqida wa-shari'a*. Cairo: Dar- al-Qalam.

Sharabasi, A.
1970 *Yas'alunaka fi al-din wa-al-hayah*, Bayrut: Dar al-Jil.

Shuqayr, N.
1916 *Tarikh Sina al-Qadim wal-Hadith wa-Geghrafiatha*. Cairo: Matba'at al-Ma'aref.

Smith, M.
1928 *Rābi'a the Mystic and her Fellow-Saints in Islam*. London: Cambridge University.

Talmon-Heller, D.
1996 "The Shaykh and the Community: The Religious Life of Hanbalite Villagers from the Nablus Region 12th-13th Centuries" in *Studies in the History of Popular Culture*, ed. B. Kedar. Jerusalem: The Zalman Shazar Center for Jewish History. pp. 95-109.

Tamari, S.
1979 "Maqam Nabi Musa near Jericho", *Cathedra* 11:153-80.
van Binsbergen, W.
1985 "The Cult of Saints in North-Western Tunisia: An Analysis of contemporary pilgrimages structures" in *Islamic Dilemma: Reformers, Naturalists and Industrialization*, ed. E. Gellner, New York, Mouton Publishers.

DOING WITH AND DOING WITHOUT: WATER USE PATTERNS AMONGST THE QASTAL FAYEZ OF BANI SAKHR

Erin Addison

Introduction

What follows is on its first level a discussion of water use patterns amongst the Qastal Fayez of the Bani Sakhr bedouin of Jordan: the essence of my observations suggests that in order to develop effective water conservation policy it is necessary to be aware in a nuanced way of the ways in which cultural history (in this case the enforced settlement of nomadic pastoralists) shapes water use. At another level, perhaps more interesting to readers of a volume such as this, the discussion of water use eventually problematizes the use of the word *bedouin* (1) to describe meaningfully a socio-cultural group in the (post-) modern Arab world, (2) to describe a marginal ("fringe") element of Arab society or one in opposition to the urban Arab world.

This account is based on many years of working and socializing daily in al-Qastal, a village on Jordan's Desert Highway, about 25 km south of Amman.[1] Al-Qastal (Fig. 1) is one of the villages of the *'ashirat* al-Fayez of Bani Sakhr, made famous as *bedouin par excellence* in Alois Musil's *Arabia Deserta* and elsewhere.[2] While it would be hazardous to generalize about all "bedouin," or nomadic pastoralists everywhere, most of the following observations apply to all of the Jordanian tribes. I was trained as an historian of religions, and my work at Qastal[3] was part of a larger effort to understand the water infrastructure that typically surrounds the Umayyad *qusur*, or "desert castles of Jordan," as they are often called. I was and still am interested in the ways in which water was used as a tool and symbol by the emerging Umayyad state. As a direct result of my work with the Fayez at Qastal, however, I returned to graduate school in 2002 to pursue a masters degree in Landscape Architecture from the University of Arizona, concentrating on arid lands issues. It seems worth noting that the material gathered as part of the process of historiography is now being pressed into the service of development policy.

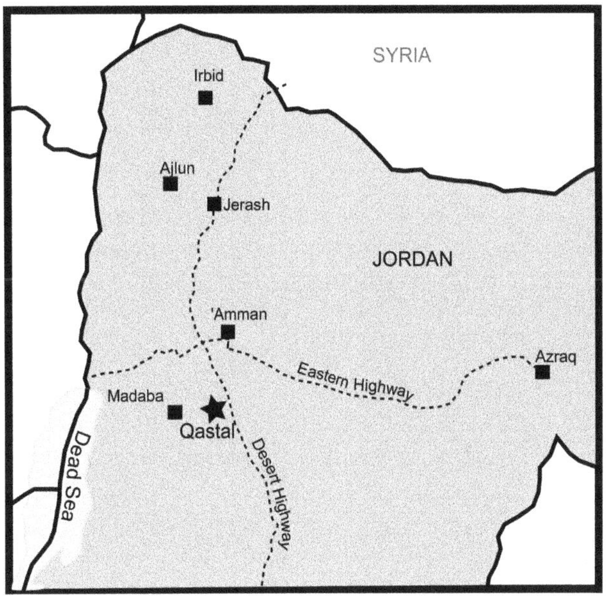

Fig. 1. Map of the region showing Qastal

Water storage and water use

The Umayyad complex at Qastal includes over 70 cisterns (the largest of which hold over 400 m^3), two large reservoirs, two dams – one over 400 meters long – and a sophisticated water harvesting infrastructure. Adjacent to Qastal – five minutes' walk from the majlis – is an iron age site called Tell az-Zubayr, where there are 18 more huge cisterns and another Umayyad reservoir. Each of the main families of the Qastal and Umm al-Amad Fayez also owns a commercial *bi'r* at Qastal – which in this case means a pump station to fill the tank trucks which supply much of Amman with its water. There is also a Water Authority *bi'r* there. In short, at Qastal water has been an important issue – a commodity worth significant investment – for at least 1,300 years and probably much, much longer.

In the process of locating and documenting the cisterns and reservoirs, particularly, we became more than intimately familiar with the Qastal community through hands-on contact with generations of their trash, which

[1] I am indebted to Fayez 'Inad Muhammad al-Fayez, 'Afash Nuri Muhammad al-Fayez, Khalid Khalid Fawwaz al-Fayez and his sons 'Ali and Khalid for, among other things, many hours of conversation about the settlement of Qastal. Notes for published sources are intended only to supplement the invaluable information I gained from my interlocutors and by observation.

[2] In what follows the word *bedouin* will refer to the historically nomadic-pastoralist tribes of southwest Asia for whom their primary residence was once the *bait shahr*, or hair tent.

[3] -- as part of the Qastal Conservation and Development Project, a non-government organization registered under the aegis of the Jordanian Ministry of Culture.

was hauled out of the cisterns by the dumptruck-load. During the second year at Qastal (1999-2000) the project transmogrified from one of purely historical interest into a project equally concerned with community development and environmental issues, and by the third year we had begun to rehabilitate some of the ancient cisterns for modern use, hoping to demonstrate the 8th century harvesting and storage techniques and use the water to sustain the landscaping planned for the site. As this work progressed and community interest in the cisterns grew, we became more and more engaged with the question of why the cisterns, if they were used at all, were used mainly for waste and why water harvesting and storage is practiced at no single house in the village.

In sum, what I will suggest is the following:

(1) the bedouin of Jordan have no cultural tradition of what I will call "water husbandry:" the harvesting and storage of water and conservative water use – they know how to "do without it," not what to "do with it;"

(2) the coerced settlement of the bedouin by the Ottoman, British and Hashemite states has been accompanied by the social and economic privileging of agriculture over pastoralism and settled land ownership over nomadic land ownership;

(3) wealthier, landed tribes have been encouraged to pursue agriculture regardless of the long-term land and water resources available to them;

(4) settlement constituted a major change in lifeways, but there has been no concomitant cultural re-tooling to address new problems – e.g. waste-management, changed means of income generation, water consumption, *et al.* – presented by settled life;

(5) water-use is attached to deep cultural symbologies, e.g., the "paradise gardens," the "fountains of The Garden (heaven)," the bedouin aesthetic and the tradition of hospitality, which encourage the luxuriant use of water;

(6) combined, these factors have made the bedouin some of the most abusive water users in Jordan.

By now it is a scholarly commonplace to acknowledge the interdependencies between the "desert and the sown," the nomadic and settled communities, which undercut the convenient dichotomy often drawn between them. But to make matters more complicated still, the Jordanian *badawin* – "bedouin," people of the *badiya*, the empty places – now by and large live their whole lives in cement houses and drive cars. In Jordanian colloquial Arabic *badawi* no longer means "*of the badiya*," it no longer actually refers to a particular lifestyle. Still, for a variety of reasons *badu* is still a politically and culturally significant self-identity in Jordan. Not the least of these is the fact that in order to consider oneself *urduni-urduni* – "Jordanian Jordanian" – one must be from one of the historically nomadic-pastoralist tribes of the Jordanian plateau or "east bank." There are farming families from the north who can trace their tenure back 500 years who are not considered truly Jordanian by tribes such as the Bani Sakhr, Huweitat or Ruwala. One need only absorb the plaintive apologetic tone of Raouf Abujaber's much-loved book, *Pioneers over Jordan,* to grasp a sense of this discourse.

The Fayez

The Fayez are an especially powerful clan within the Bani Sakhr[4] – at the time of writing (summer 2004) Jordan's Prime Minister is Faisal al-Fayez from Umm al-Amad, a village nearby Qastal – but they are by no means the only politically and economically enfranchised bedouin in Jordan. Indeed the Hashemite kings continue to style themselves as bedouin sheikhs and much of the business of governance is conducted in traditional tribal manner: the "sheikhs" – the Hashemite royal family – sit in *majlis,* receive clients who have a greater or lesser claim to their attention based on their tribal affinities, and grant favors according to much the same criterion. In Jordan today, to identify one's lineage as "bedouin" confers a particular nationalist legitimacy and primacy, and connects one to a vast network of intra-tribal relationships and benefits. Thus it is difficult to configure the bedouin as the "fringe" of Jordanian society – the tribes are its very core.

According to Abujaber, Sattam Fandi 'Abbas al-Fayez in the 1870s acquired a vast tract of land – which included al-Qastal – from sheikh Diyab al-'Adwan of the Balqa' (180-181). The Bani

Sakhr pride themselves on their heritage as one of the great camel herding tribes, and are known for their horses, their record of military service, and their loyalty to King Hussain (Fayez, *passim*). The Fayez, however, are also notable at least partly for the fact that Sattam was the first of the great bedouin sheikhs to lay hand to plough and make a significant enterprise of agriculture (Abujaber 177-196). Muhammad Fawwaz Sattam built a house on top of the hill at Qastal in 1954 and the family of Khalid Fawwaz Sattam moved into the Ottoman police station at Qastal[5] sometime after the First World War. As the Ottoman police station had already been constructed, *bait* Sattam was probably the first building in Qastal constructed from scratch, expressly for the purpose of settled habitation, after the Umayyad period.

[4] Three Fayiz sheiks served as regional governors during the Ottoman period (Abujaber 1989) and a dozen more served as high-ranking officials in the Hashemite armies, the court and Jordanian government (Fayez, *passim*). The Bani Sakhr includes the Khreisheh and Zubn clans, who have also been politically influential.

[5] Based on the pottery evidence and the inscriptions on the headstones in the ancient graveyard, the Umayyad palace at Qastal was used continuously through the Mameluke period. In the Ottoman period the northwest quadrant of the palace was refitted as a police post, into which Khalid Fawwaz Sattam moved sometime after the First World War.

Fig. 2. Water truck and sheep ready for transport (photo by the author)

Fig. 3. Modern farm (photo by the author)

Five generations later, like most of Jordan's population, the Fayez are settled bedouin. Qastal is a largish village with an industrial estate on it and thousands of dunums of olive orchards and greenhouses. Today, if they have livestock herds at all, Syrians, Iraqis and Palestinians herd them – and often drive them home in the evening in trucks to corrals (Fig. 2). The Qastal Fayez are still, in a new way, semi-nomadic – the more affluent typically have a house on the farm (Fig. 3), a house in their home village (in this case Qastal), and a house in the fashionable suburbs of West 'Amman. They buy decorative paisley tents made in India and go out to the *badiya* for a couple of weeks in the spring to shoot things up and drink some Black Label. Nestled within this change from the empty places to cement villages is, I think, the most important single issue in understanding water use in these communities, namely, that the shift to settled existence and the consumption habits that come from it have not been accompanied by concomitant changes in approaches to natural resources (Sanlaville 13-14; Velud 74ff.; Bocco 209 n.44).

Before work at Qastal began we were aware from earlier research (Carlier & Morin 1985, 1986; Carlier 1984, 1989) that there existed abundant cisterns onsite and that the qasr had a huge central cistern under its courtyard, capable of holding as much as 1,200m^3. When we began work in 1998, however, the enormous extent of the system was unknown, the cisterns had not been mapped or documented and no connection had been established between the surrounding reservoirs and dams, the cisterns, and the Umayyad complex. Furthermore, in the late 1980s and early 1990s the antiquities site had been used as a dump for debris from the construction of the frontage road, and after that as a town dump. The huge central cistern had for some years been closed and used as a septic tank by *dar* Shibli Khalid Fawwaz al-Fayez.

The cisterns

As the debris was cleared, we started to uncover the mouths to cisterns and mark them. Most weeks one or two of the local Fayez patriarchs walked through the site to watch the project, and eventually they began to point out places where they remembered there to have been cisterns. Without exception, if we dug where they showed us, indeed there was a cistern. Though cisterns were inevitably filled with debris and closed with blocks from the qasr, it was clear that they were of some significance to the older generation.

We eventually determined that the cisterns at Qastal had once – during the Umayyad period – been connected to each other, to a spring on the hill, and to roof harvesting at the qasr. Though there is ample evidence that surface catchment was important to the original Umayyad system, the cisterns are capable of holding far more than the evident surfaces could have harvested, and we determined that the linked cisterns stored water from a spring on a hill in middle of town. The spring on the hill had run dry in the mid-80s, but even before that the Fayez had not utilized the cisterns to store water – households carried water from the spring. Because Qastal lies within the 300mm rainfall line, barley was dry-farmed. Eventually we discovered that before the Fayez were completely established in Qastal, the qasr had been used as stabling and the cisterns had been used to store *tibn*, or dry fodder for livestock. The cisterns which had never been used for trash – only for tibn – would turn out to be in very nearly perfect repair, and still capable of holding water. The very fact that hay could be stored in them for over two years without being damaged by moisture suggests the integrity of the cistern linings. In sum, water was always scarce at Qastal, elaborate infrastucture for harvesting and storage was available, some of it still functional, but we encountered only one cistern – of dozens – that had in fact been used to store water, and it was no longer in use.

Within a generation of settlement, piped water was installed in some houses in Qastal and the commercial wells began to operate, making cheap trucked water available to anyone who could afford JD50 for a tank on his roof. Moreover, those in the best position to invest in water infrastructure and the greatest need for water for irrigation – the principal Fayez households at Qastal – owned the commercial wells.[6]

Water use

At first glance Qastal appears to be a very water-thrifty community. But a closer look at the high-income homes reveals an often gross waste of the resource. The garden aesthetic is entwined with a deep ethic of hospitality, and the paradigmatic garden is green, deeply shaded and includes the sound of running water and hardscape on which to place chairs and tables.

Created landscapes tend toward traditional stereotypes which prefer exotic species, high-water-use plantings and large water features (Addison & Livingston, 2004). Choices of landscaping

plants adhere to these aesthetics with little regard for water resources. Government water is piped in for 8-12 hours once or twice a week depending on the season, and during those hours *dar* Shibli and others let hoses run open on their gardens for the entire period that the water is coming. Household water use also tends away from conservation. There exists a purity ethic which prefers cleaning with water over "dry cleaning", i.e., sweeping or dusting, and the more water is used the cleaner things – including expanses of hardscape – are thought to be. Crops are chosen purely with regard to market demand and not to their water needs, since for the largest producers – the Qastal Fayez – water is all but free. In sum, those who can waste water do so. Lower income houses exhibit the same use patterns on a smaller scale.

Bedouin are often romanticized as a people who lived in harmony with nature, as innate conservationists.[7] In fact this is belied by most of our experience in Jordan. The bedouin are enormously *resourceful* within a context of scarce natural resources. They are skilled at using what comes to hand. Østen LaBianca talks about "indigenous hardiness structures", which include the tendency toward low-overhead, low-investment projects (1997),

[6] In a 1974 report by the Jordanian Department of Natural Resources – only twenty years after the construction of *bait* Muhammad – no fewer than seven groundwater wells are noted for Qastal municipality itself (Abu Ajamieh 42, fig. 2)

[7] For an extreme example see Rami Sajdi's discussion of the 'Amareen tribe of southern Jordan at www.bedouincamp.net.

maintaining flexibility, and the ultimate tendency to decamp when resources run out. Before pasturing circuits were obstructed by national boundaries, when life-expectancy was limited and infant mortality rates high, and before urbanization and the extension of agriculture into the *badiya* depleted the water table, the Bani Sakhr did, in fact, live more or less in harmony with their resources: they moved from water source to water source over a relatively large expanse of territory, depleting the resources at a site and moving on, effectively letting that land "lie fallow" until the next year (Abujaber 290 n.44). Improvement of water sources, however, was minimal, as we can see from countless sites throughout the badiya, most of which have not been significantly improved since Umayyad or even Nabataean times. Thus the Bani Sakhr, at least, have no cultural tradition of water husbandry – they were skilled at doing without water. Now that water is a resource seemingly abundant and immediately at hand, they are using it essentially as a low-overhead crop.

Numberless personal discussions with local residents at Qastal as well as the owners of the commercial wells has yielded ample evidence that the relationships between pumping, drought, rainfall, springs, and the water table is ill-comprehended. Again and again we encountered certain common, prevailing misconceptions:

(1) that the drying of springs and lowering of the water level in wells was due to rainless years; in other words, that the lowering of the water table was solely drought-related;

(2) that rain would restore the springs and wells; in the rare cases in which the interviewee was aware of the existence of the aquifer, they believed it rechargeable in the short term, e.g., after three or four good rain years;

(3) that should Jordan actually run out of water, money and technology could drill deeper for or pipe water over immeasurable distances (e.g., from Turkey and even Libya!).

Thus ignorance about water supply and its mechanics reinforce the abuse of resources by sustaining the impression of an indefinite supply of water. Water itself is "harvested" without foresight for the effects of pumping on agriculture or potable water supply.

An important piece of this picture is the fact that the bedouin were either forced or actively encouraged – both by the state and by the institution of national borders – to settle (Hiatt, *passim*). Moreover, there has been a persistent effort to engage them in agriculture, and agriculture has been privileged by the state in terms of development input and taxation (Tall, 90ff.). There continues to be a concerted effort to extend irrigated agriculture further and further into the badiya. Even newly developed water sources – e.g., treated wastewater – often tend to be viewed as surplus water, not as a substitute for pumped groundwater. At the same time little effort has been made to adapt bedouin cultural patterns to a new lifeway and livelihood. The latter has adduced problems not only of water management, but income generation and waste management as well. Three generations ago, bedouin did not carry significant amounts of non-biodegradable materials such as plastics and polystyrene. As long as they kept moving, they left manageable amounts of biodegradable waste behind them, and their consumption of hard goods was limited by their mobility. Nomadic pastoralists never developed a tradition of waste management simply because it was never necessary.

Technology and population increase have complicated both waste management and water harvesting. In the case of semi-nomadic groups, waste management has been exacerbated by rising population and the development of affordable, non-biodegradable products. In the case of nomadic groups, trucks have made it more feasible to carry lightweight non-biodegradable materials. Thus waste materials are produced, but no management system has been introduced at the same time. Similarly, tank trucks now move water into the *badiya*, which means that the very few remaining nomadic groups have not had to develop new harvesting methods. Instead of moving from water source to water source, the water moves to them.

The same interlocking dynamics apply to formerly nomadic groups who are now settled: population increase, increase in the use of non-biodegradable goods, and the development of water technology. In the case of settled groups these dynamics are intensified by static settlement. Mobility no longer limits the consumption of non-biodegradable goods. Biodegradable waste is accumulated in place faster than it can degrade. Because water is now affordably piped or transported in tank trucks to residences, waste management has been a more pressing problem than water. Water is brought in trucks and stored in tanks on roofs, but roofs and terraces are not used as catchment surfaces to harvest rainwater to store in cisterns which already exist. Lacking a tradition of water harvesting and the use of cisterns, settled bedouin communities have used the cisterns for waste disposal.

Clearly the problem is *not* that formerly bedouin communities *cannot* learn new technologies quickly and effectively. At Qastal the Fayez pump water in enormous quantities at Bi'r Akif and Bi'r Humeidi for sale on the commercial market. They also operate some of the most sophisticated olive presses in the country and employ efficient drip irrigation on many olive farms. Yet waste from the presses is crudely disposed of in cisterns. To fall back on folk wisdom, however, "necessity is the mother of invention". It hasn't been necessary for the Fayez to be very inventive about water. Making a living and disposing of waste have been pressing and immediate needs, whereas water has been made easily available: there has not been a perceived need to learn new – or old – water technologies – especially ones which require considerable overhead in terms of labor and construction.

From an immediate policy point of view, the mechanics of water abuse by the Fayez reveals the importance of understanding both recent history and the *longue duree* if we are to implement conservation practices effectively at all: clearly education is an urgent priority need; waste management and population control, among others, have to be implemented alongside water management measures; xeriscape measures will have to be adapted to a complex aesthetic tradition with important social and metaphysical associations; water needs to be rationed and, beyond the subsistence level, taxed aggressively to discourage commercial pumping and encourage water harvesting which does not stress the aquifers.

As for the subject of the bedouin, *per se*, it would seem that the foregoing problematizes the word altogether. Water use is one set of concrete practices which has changed dramatically within in a few generations, even as more conceptual expressions of bedouin identity persist. The investigation just of this one particular aspect of transition in bedouin life does much to disabuse the romantic illusions about the bedouin's relationship to the natural landscape they inhabit, and certainly to dismantle any kind of existing dichotomy at all between bedu and fellah. The designation *bedouin* in Jordan is today almost entirely disconnected from nomadic-pastoralism except in a strictly historical sense – and yet the word still conveys a powerful political and cultural self-identity. It seems to this author that a careful dissection of the transitions effected during state-supported and state-enforced settlement may reveal the very basis of what constitutes bedouin identity.

Acknowledgements

The author would like to acknowledge Fulbright Hays Group Research Abroad, The Fulbright Foundation, & the Small Grants Programme of the United Nations Development Programme (UNDP) Global Environment Fund for their generous support of the work at Qastal from 1998-2002; the International Arid Lands Consortium and the University of Arizona School of Landscape Architecture for their support of my current research on water conservation education in Jordan and their support for travel to the symposium for which this paper was written.

Bibliography

Abu Ajamieh, M.
1974 *The Use of Gravity Groundwater Prospecting in Zyzya Area.* Amman: Department of Natural Resources.
Abujaber, R.S.
1989 *Pioneers Over Jordan.* London: Tauris.
Addison, E., and Livingston, M.
f.c. Cultural Interpretation and the Implementation of Water Conservation Policy. *Proceedings of the Dead Sea Conference on International Water Demand Management.* Amman.
Bocco, R.
2000 "International organisations and the settlement of nomads in the Arab Middle East, 1950-1990" in *The Transformation of Nomadic Society in the Arab East*, ed. M. Mundy and B. Musallam. Cambridge: Cambridge University. pp 197-217.
Carlier, P.
1984 "Recherches Archaeologiques au chateau de Qastal (Jordanie)". *Annual of the Department of Antiquities of Jordan* 18:353-383.
1989 "Qastal al-Balqa': an Umayyad Site in Jordan". in *Proceedings of the Fourth International Conference on the History of Bilad ash-Sham*, ed. M. A. Bahit and R. Schick. Amman: University of Jordan. pp. 104-39.
Carlier, P., and Morin, F.
1985 *Archaeological Researches at Qastal '85: second campaign preliminary report.* Report filed at the American Center for Oriental Research, Amman.
1986 *Qastal al-Balqa.* Grignan: publ
Fayez, M.'A. S.
1995 *'Asha'ir bani sakhr: tarikh wa-muwaqaf hata sanna 1950.* Amman: University of Jordan.
Hiatt, J. M.
1981 *Between Desert and Town: a Case Study of Encapsulation and Sedentarization among Jordanian Bedouin.* Unpublished Masters thesis. Ann Arbor, MI.
La Bianca, Ø.S.
1997 "Indigenous hardiness structures and state formation in Jordan: Towards a history of Jordan's resident Arab population" in *The third Nordic conference on Middle Eastern Studies: Ethnic encounter and culture change, Joensuu, Finland, 19-22 June 1995*, eds Sabour, M. and Vikør, K.S. Bergen/ London: Hurst. pp 143-57.
Musil, A.
1907 *Arabia Petraea, Vol. I: Moab.* Vienna: Kaiserliche Akademie der Wissenschaften.
1927 *Arabia Deserta.* New York: American Geographical Society.
Sajdi, R.
2004 The Fuqara (Shamans) of the Ammarin, *et al.* www.bedouincamp.net. Accessed 12 June 2004.
Sanlaville, P.
2000 "Environment and development" in *The Transformation of Nomadic Society in the Arab East*, ed. M. Mundy and B. Musallam. Cambridge: Cambridge University. pp 6-16.
Tall, T.
2000 "The politics of rural policy in East Jordan, 1920-1989" in *The Transformation of Nomadic Society in the Arab East*, ed. M. Mundy and B. Musallam. Cambridge: Cambridge University. pp 90-98.
Velud, Ch.
2000 "French Mandate policy in the Syrian steppe" in *The Transformation of Nomadic Society in the Arab East*, ed. M. Mundy and B. Musallam. Cambridge: Cambridge University. pp 63-81.

BEDOUIN CULTURAL REMAINS IN THE EILAT REGION

Uzi Avner

Introduction

Desert societies of the past, and the desert as a human habitat, were often viewed negatively by scholars. Desert people were described as nomads roaming the area, destroying existing cultural remains and leaving no traces of their own (Glueck 1935: 183; 1968: 11-12, 127; 1970: 65), "*Always hungry, and unceasingly they cast longing eyes upon lands which to them are lands flowing with milk and honey*" (Glueck 1970: 11). The desert "*could not sustain a local population for any length of time*" (Haiman 1992a: 93), desert societies "*could not have existed without the support of a strong stable political and economic body*" (Beit-Arieh 1984: 22). The desert population was always "*hungry, on the verge of death*" (Haiman 1992b: 304). "*In the desert areas... groups that practice subsistence economy based on hunting and gathering or animal husbandry... do not leave traceable remains*" (Finkelstein and Perevolotsky 1990: 68).

These few quotations represent a broader misconception of the desert environment and populations, and necessitate brief preliminary comments. First, both ancient societies and the more recent bedouin in the deserts of the Southern Levant were never nomads but only semi-nomads, and the cultural differences between the two are great.[1] Second, even true nomads do leave remains (*e.g.* Cribb 1991), and certainly so semi-nomads, including the bedouin (*e.g.* Avner 1990; Avni 1992; Banning, and Köller-Rollefson 1992; Eldar *et al.* 1992; Goren-Inbar 1993; Saidel 2001). A closer look into the bedouin culture of the Eilat region may shed some light on desert life.

The Haywat

Over the last five centuries, the Eilat region was the domain of the *Haywat* (or *Ahaywat*), a small tribe (more accurately a confederation of three small tribes) which numbered some 1000 people before 1948 (Marx 1967: 17). The precise boundaries of the tribe's territory are somewhat questionable. An unusual document written by the tribe's members in 1969 and studied by Stewart (1986) presents their territorial claim, *ca.* 3500 sq km (Fig. 1). In reality, their territory is smaller since the southern part is occupied by the southern branch of the *Tarabin*, and the north was invaded by the *Tiyaha* confederation (the *'Azazme, Shegeirat* and *Breykat*). In the 16th century their territory still included parts of northern Hejjaz, south of 'Aqaba (Stewart 1986: 18, with references), and in the early 20th century they still made pilgrimages to a holy tomb a three day walk south of 'Aqaba ('Aref al'Aref 1937: 187). During the 1948 war, families who lived in the Eilat region, between the Egyptian and the Jordanian borders, moved to Southern Jordan, and some 150 families who lived in Sinai remained disconnected from the rest of the tribe. Today, Kibbutz Yotvata maintains contact with *Haywat* members in Jordan.

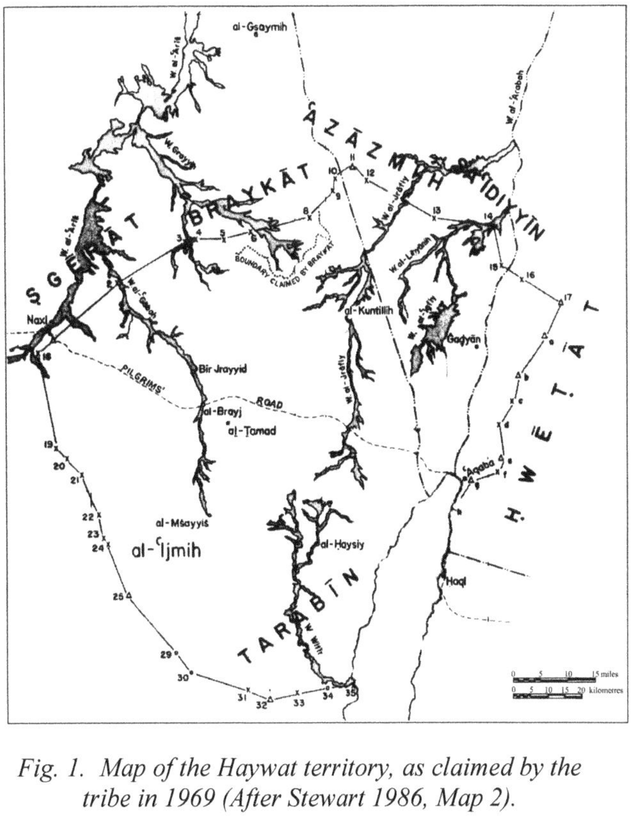

Fig. 1. Map of the Haywat territory, as claimed by the tribe in 1969 (After Stewart 1986, Map 2).

[1] Full nomads subsist on herding alone, and migrate hundreds of kilometers twice a year. In the Negev and Sinai, both ancient populations and recent Bedouin migrated with the herds only tens of kilometers; they subsisted on a complex economy, which included agriculture and several crafts (Marx 1967; Abu-Rabia' 1994). They sometimes altered their mode of life, whether towards more mobility or more sedentism, in response to the political situation or to climate changes (Marx 1992, 1996; Khazanov & Bar- Yosef 1993; Finkelstein 1995).

Fig. 3. The Bedouin palm grove at Yotvata ('Ein Ghadhian), on a rare foggy morning

Fig. 2. The water well of Kuntila, 70 km NW of Eilat

The *Haywat* territory is hyper-arid, with an average annual rainfall of only 25-30 mm, in contrast to *ca.* 4000 mm of annual potential evaporation. As a result, vegetation is restricted to wadi beds and the carrying capacity is low, and most of the area can only be exploited for goat and camel herding. Nevertheless, their territory contained several important assets. One was the water well of Kuntila (Fig. 2), some 70 km northwest of 'Aqaba, adjacent to Darb Ghaza, the major ancient road connecting the Red Sea with the Mediterranean. The water was pure and plentiful, and passing caravans or travelers were obliged to purchase it to replenish their supplies. Another was Wadi 'Uqfi (today 'Uvda Valley) where the tribe was very successful in growing barley and wheat (see below). The third asset was the oasis of 'Ein Ghadhian (today Yotvata), an important water source on the 'Arabah road, with a palm grove (Fig. 3).

The tribe had at least 26 water wells in its territory[2] and many natural, seasonal cisterns that retained water from one to four months, if a flood occurred (Fig. 4). There were also six small springs, nine *thmeilat* (excavated water holes in wadi gravel) and other water sources.

Fig. 4. Bedouin watering camels in Muyat Galla, west of Jebel Hashem alTaref

Another important property was Darb al-Hajj, the Muslim pilgrim road from Cairo to Mecca. The road crossed the *Haywat* territory in the section from Nakhel to 'Aqaba (some 140 km), and according to centuries old customs, they held the right to escort the Hajj caravans through their section of the road (Fig. 5), and were paid for it by the government (Mamlukian or Ottoman). However, in the 16th century, soon after the 'Ottoman conquest, the *Haywat* lost control of the most sensitive section of the road, the section from Ras al-Naqeb to 'Aqaba, to the *'Alawin* of the *Hwetat* confederation, their neighbors from the east. They also lost control of parts of

[2] Musil (1907-8, III: 45-46) listed 11 water wells in the *Haywat* territory. Jennings-Bramley listed 17 water wells in the same area, in an unpublished document written between 1904 and 1907, quoted by Stewart (1986: 36). Both lists are incomplete, and I could add at least 11 more wells and other water sources.

Fig. 5. Hajj caravan escorted by Bedouin, as seen by W. H. Bartlett in 1845, west of Qa' alNaqeb (Bartlett 1849:152), drawn by C. Cousen)

the 'Arabah roads and the roads leading to St. Catherine,[3] which were still within their territory. In 1883, the British mandate discontinued the annual Hajj caravan through Sinai (Winter 1987:801), and the tribe lost this source of income. Another indication of the tribe's weakness was that during the 19th century, explorers traveling the major roads crossing their territory were escorted by other tribesmen, not by *Haywat*.[4]

The *Haywat* rarely attracted the attention of explorers and scholars. Among the early scholars, Alois Musil was almost the only one who referred to the tribe. He was guided by *Haywat* members in the Eilat region and the southern 'Arabah (Musil 1907-8 II/2:178ff), and in his thorough account of the bedouin ethnography (1907-8 III) he frequently, although briefly, addressed the *Haywat* (among the other tribes), especially in poetry. He also listed the tribe's water wells (see Note 2). Jennings-Bramley explored Sinai in the very early 20th century, particularly the *Haywat* territory. He published a series of reports (1905-1914), but only rarely mentioned the tribe (*e.g.* 1905:127; 1906:25; 1908:30f). He did refer to their borders (ibid), and in an additional, unpublished document he listed their water sources (see Note 2). Woolley and Lawrence (1915) explored the Negev and part of Sinai, but did not mention the *Haywat*.[5] Major C. S. Jarvis, the British governor of Sinai after World War I, wrote a series of books on the peninsula, but in the three most important volumes (1931, 1936, 1938) did not address the tribe. 'Aref al 'Aref, the governor of the Negev after World War I, wrote two books on the Negev bedouin, but mentioned the *Haywat* only once, and in passing (see above). Fritz Frank surveyed the 'Arabah Valley and the Eilat region in 1932 and 1933, guided by *Haywat* members (Frank 1934:246, 248, 261, 266, and see here Fig 16 and Note 7). He referred to the *Haywat* borders in two places (pp. 233, 266) but provided no further information about the tribe. During the 1970s, Frank Stewart thoroughly studied the *Haywat*'s legal system and published a series of books (Stewart. 1977, 1986, 1988, 1994, 2003), but did not relate to other aspects of the tribe's culture.[6] The *Haywat* were looked down upon by neighboring tribes; *Tarabin* members in Nuweiba' and Wadi Watir called them "lizard eaters" (M. Shemtov, *pers. comm.*).

In light of the harshness of their land, their low attraction for scholars and their poor present situation, one would not expect to find many remains of the *Haywat* culture. In fact, this impression is inaccurate, as is demonstrated by several sources. One can be found in the Eilat region, where they lived until 1948, and left many "archaeological" remains: tent camps (*atlal,* Fig. 6), with scattered sherds of Gaza Ware, and where millstones, wooden bowls and complete jars are occasionally found as well (Figs 7, 15); storage places built in rock shelters (*mahzan* or *matmara*, Fig. 8), some still containing various utensils; corrals of various types with layers of animal dung (Fig. 9); open-air herd stationing places (Fig. 10); predator traps built to protect the herds; open-air mosques (Fig. 11) and large

[3] Al Jaziri, mid 16th century, quoted by Stewart 1986:22-23; see also Burckhardt 1822: 512.

[4] Burckhardt 1822: 438-440; Rüppell 1829:215, 241; Robinson 1856: 166-168, 173, and see further examples in Stewart 1986: 24-24. In 1845, W. H. Bartlett met with a Hajj caravan west of Ras al-Naqeb and described it vividly (1849: 149-162, and see Fig. 5). On his way from 'Aqaba to Cairo he was escorted by *Hwetat* men, not *Haywat*, as far as Nahel, and there he was obliged to take *Tiyaha* guides (p. 163).

[5] The *Haywat* is not mentioned either in the other books of T. E. Lawrence, "Revolt in the Desert" (1927) and "Seven Pillars of Wisdom" (1935).

[6] Two studies (Murray 1935: 248-251, and Oppenheim 1939-68 Vol 2: 149-153) did mention the *Haywat*, but Stewart (1989: 2, N. 8) noted that they actually addressed the *Safayhih Ahywat*, a different, separate group.

Fig. 6. Nahal Girzi (Wadi alGhabiyeh), the remains of rectangular Bedouin tent

Fig. 7. Pottery from various Bedouin camp sites north of Eilat: 1, 2, 3, 6- Ghaza Ware; 4- 'Othoman, 5- Mamlukian, the latter two were still in use until 1948. No. 1 found by M. Shemtov, No. 2 by El'ad Saragusti, 3-6- by the writer.

Fig. 8. Bedouin storage near Be'er Ora (Bir Hindis).

Fig. 9. Eastern 'Uvda Valley, an ancient corral in a rock-shelter, reused by Bedouin.

Fig. 10. Nahal Tsihor (Wadi Suhair), an open-air herd stationing site.

Fig. 11. Wadi Hessy, a small open-air mosque; the niche (mihrab) points southward, towards Mecca.

and small burial grounds (*maqbara*, Fig. 12). Another source of interest is provided by encounters with the bedouin in Sinai during excursions or archaeological work (1969-1982), and by observing their way of life. No comprehensive ethnographic work has ever been undertaken on the *Haywat;* however, several of my friends maintained close contacts with tribesmen, and some collected or purchased their old artifacts and utensils.

Bedouin Property and Utensils

Despite the environmental conditions, the *Haywat* raised fairly large herds of goats and camels (Figs 4, 13), which served as their basic and most stable source of subsistence. In fact, the wide and shallow wadis support quite a rich vegetation, with over 40 plant species edible for animals. The density of corrals and open-air herd stationing sites (Figs 9, 10) indicates the importance of herding for the tribe's economy. This is true for every bedouin society, since both camels and goats are highly adapted to harsh desert conditions (Shkolnik 1977). They are very efficient in producing food for people, both storable milk products and meat. During years with rich pasture, bedouin enlarge their herds as much as possible in order to trade the animals for other products in city markets, and reserve others as an "insurance" for lean years. Flexibility in herd size is part of the survival strategy of the bedouin.[7]

Fig. 12. Yotvata ('Ein Ghadhian), part of the central burial ground of the Haywat *(c.f. Musil 1907-8 III:45). Notice the many tombstones in the photo's centre*

Fig. 13. Qa' alNaqeb, camels and goats of the Haywat.

Fig. 14. 'Awde Salame, the guide of F. Frank in 1932-3, in his tent in Wadi Tweibeh, 1971.

When visiting a tent camp, one realizes that the bedouin hold a considerable amount and variety of facilities and utensils (Fig. 14).[8] Examples are presented in Figs 15-22, with brief explanations in the captions. The majority of items presented here was photographed from the collection of Alfonso Nussbaumer, Eilat, for the sake of this paper. Many of them have now disappeared, and been replaced by modern substitutes. Although they represent only part of their material culture, these objects do illuminate the bedouin traditional way of life. Basketry, wood, leatherworks and weaving were always part of their everyday practice. Pottery and metal objects were acquired through trade; however, all utensils, including metalworks, were repeatedly repaired by the bedouin, and reused.

Fig. 15. Wooden utensils: 1. maswata, *for milking animals; 2.* 'umra, *for storing small cups* (fanagin); *3.* batiya, *for preparation and surving food. 4.* hanaba, *for preparation and serving food.*

[7] Frank (1934:246) noted that in previous years the family of 'Awde Salame, his guide, raised hundreds of goats, but when he met them they had only several dozens, due to a series of lean years.

[8] The old man in Fig. 14 is 'Awde (='Ode) Salame, the guide of Fritz Frank in 1932-3. When the photograph was taken, in 1970 in Wadi Tweibeh, he was blind, but his knowledge and memory of the area was phenomenal. He died in 1974.

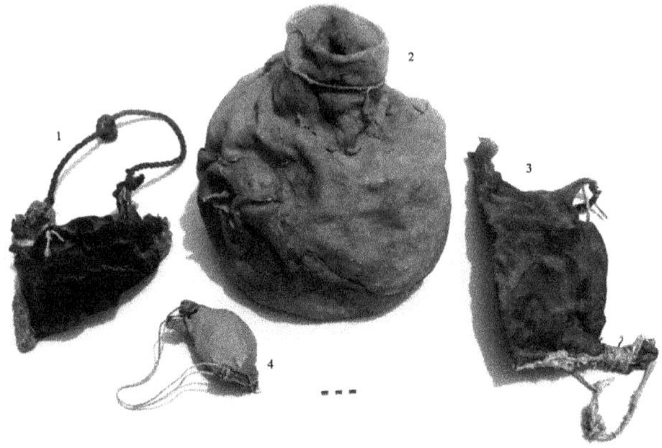

Fig. 16. Leather utensils: 1. girbeh, *a water skin; 2.* balla/gar'a, *made of camel stomach, for storage of crushed dates or for butter, 3.* si'an, *for preparation of milk products; 4.* garab, *made from a lizard skin for storage of tea, coffee and others.*

Fig. 19. Coffee and tea utensils: 1. pestle (mehbas/sahbut); *2. mortar* (hawan/gurn*); 3. frying pen for coffee beans* (mahmasa); *4, cup box* ('umra*); 5. coffee pot* (bakraj*); 6. tea pot* (dalla/ ibirq*). 7. tea and coffee bag* (garab); *8. coffee cup* (finjan); *9. tea cup* (qubayeh)

Fig. 17. Basketry of palm leaf and palm fibers: 1. begs (kees); 2. chicken cage; 3. sandals.

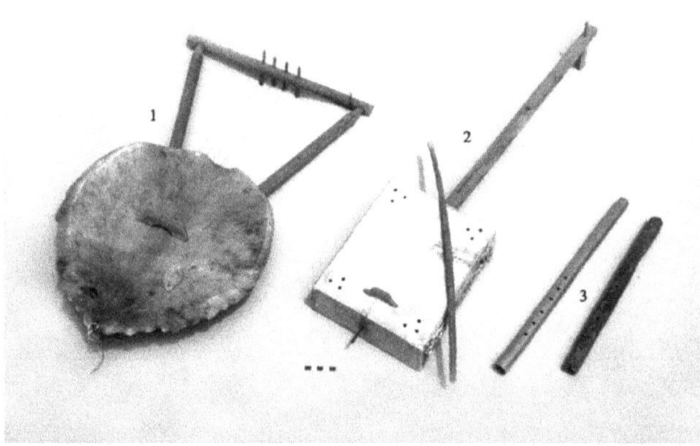

Fig. 20. Musical instruments: 1. sumsumiya, *made of see turtle shell; 2.* rababa *(single string violine); 3.* shubab, *flutes improvised from metal tubes*

Fig. 18. Copper cooking pots (kider). *The bottom of the larger, used for communal meals, was replaced three times*

Fig. 21. Wooden combs from tent camp remains near Be'er Milhan (probably originated in Syria)

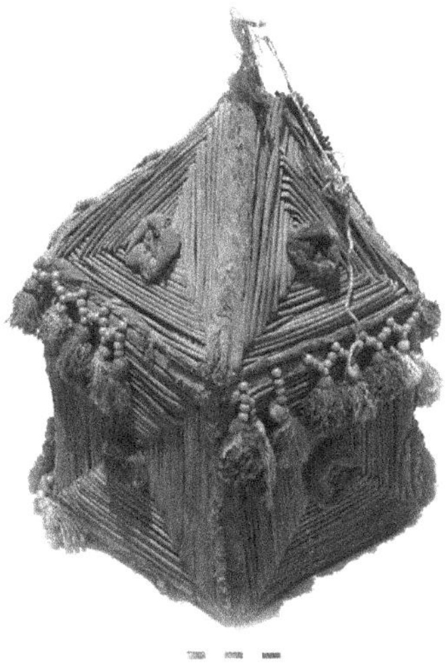

Fig. 22. Woman's Jewellery box (qutta)

Fig. 23. Weaving of sheep wool (quite rare among the Haywat*), near Kuntila.*

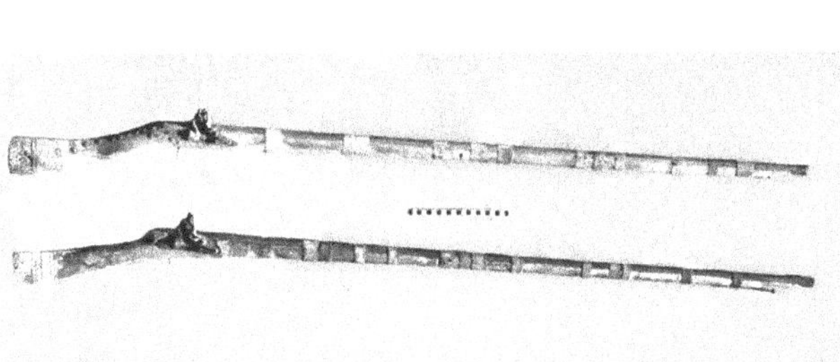

Fig. 24. Turkish firearms from the 18th century (barud), found by G. Yahalom and G. levinstein in Nahal Roded.

Fig. 25. Gun powder containers (qasid), *made of wood (left) and copper (right).*

Fig. 26. Eastern 'Uvda Valley, a cave storage of agricultural implement, including a wooden plow beam, iron plow tip, two iron sickles, millstones and others.

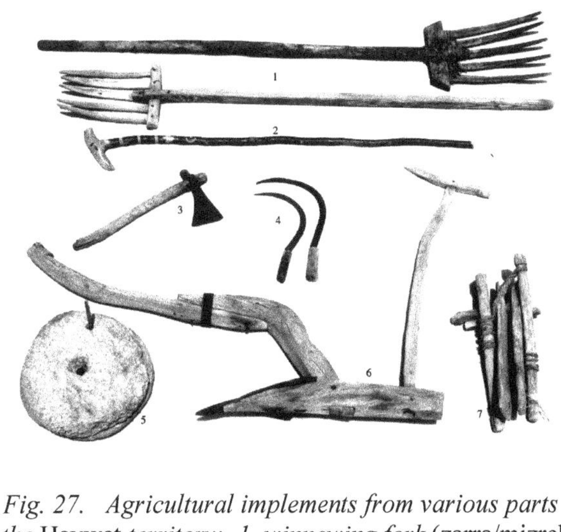

Fig. 27. Agricultural implements from various parts of the Haywat *territory: 1. winnowing fork* (zarra/mizreh); *2. stick for cleaning the plow and for picking fruits* (minfis); *3. an axe* (shareq); *4. sickles* (mahesh); *5. millstones* (raha); *6. plow* (mahrash). *7.* jadem, *a saddle for carrying wheat and barley yields on a donkey or camel back (collapsed).*

Bedouin Agriculture

While travelling through *Haywat* land, one can often see ancient remains of cultivated fields in the wide and shallow wadis, with low terraces or embankments. Some are currently cultivated by the tribe, especially around Nakhel and Themed, in the upper Wadi Hessy and in 'Uvda Valley.

'Uvda Valley (*Wadi 'Uqfi*, or *'Igfi* in bedouin pronunciation), 40 km north of Eilat, was the most important agricultural area. On the eastern side of the valley the remains of a vast ancient agriculture settlement system, which began in the 6th millennium B.C, has been studied since 1978 (*e.g.* Avner, 1998, 2002a, b; Avner *et al.* 2003). During the archaeological survey and excavations we also found many indications of bedouin agriculture. Several caves and rock shelter storages contained agricultural implements, such as iron sickles and hoes, wooden plow beams, iron plow tips, millstones and the like (Fig. 26). Similar equipment is occasionally found in other locations in the Eilat region.[9] Hundreds of bedouin silos are still visible today on the surface (see below), and even plow marks are still visible.[10]

Short, but important descriptions of bedouin cultivating the valley, were provided by several explorers. Musil, who visited the area in 1902, mentioned it while referring to the people of 'Aqaba: "*In Wadi 'Uqfi, the soil is good for cultivation and the crop is plentiful when the rain is sufficient to create floods. In this valley the citizens* (of 'Aqaba) *are renting plots from the Haiwat, sowing wheat and barley and living in tents by the fields in the seasons of sowing and harvesting. After the harvest they return to 'Aqaba with the threshed grain"* (Musil 1926:85).

On December 29, 1979, I visited the valley with four *Haywat* elders who had cultivated the land in the past, in order to learn about their agricultural practices. The visit was undertaken with the assistance of anthropologist F. Stewart, who was at the time studying the tribe's legal system. Following is the essential information supplied by the elders:

1. Between 1939-1948, the *Haywat* were able to cultivate the valley four times, *i.e.* almost every second year, when the amount of rain was sufficient to create floods. However, nearly every year of sowing provided a good yield (see below).

Fig. 28. Eastern 'Uvda Valley, an ancient threshing floor reused by the Haywat *until 1948.*

Fig. 29. Western Negev Highlands, Bedouin threshing barley by animal trampling (photo by M. Haiman, 1981).

2. Approximately 20 days after a flood, while the soil was still moist but not too wet, they sowed barley and wheat by hand. Then they plowed the soil in order to cover the seeds, using light wooden plows pulled by camels.
3. Harvest took place in the late spring-early summer (May-June), using iron sickles, and the threshing season lasted until mid- or late summer (July and August). This was done either on temporary threshing floors in the fields or on the ancient ones, most of which were cut into the rock surface; the *Haywat* claimed the latter were built by their own fathers. In fact, we did find bedouin remains in most ancient threshing floors.[11] They threshed by animals trampling, as it was practiced in other parts of the Negev and Sinai (Fig. 29), while the threshing sledge was unknown to them.
4. According to the elders, the yields in the 'Uvda Valley were the highest in their territory, around 800 kg per hectare (translated from their own measurements of weight and area). This enabled them to lease out plots to citizens of 'Aqaba, in return for a

[9] In August 2004. M. Shemtov found deteriorating wooden plows and other agricultural gear in a rock shelter storage near Be'er Ora, and others were previously found by the writer northeast of Timna', both in the 'Arabah Valley. No plots cultivated by the Bedouin are known in the southern 'Arabah, besides those near 'Ein Ghadhian (Yotvata) and 'Aqaba; however, Qa' al Sa'idin, 60 km north of 'Aqaba, was cultivated by the *Sa'idin* tribe.

[10] Stewart (1986:21, N. 31) quoted *Haywat* members saying that in the past 'Uvda Valley was the property of St. Katherine's monastery. Burckhardt (1822:546-548) heard in the monastery of Cairo that all Sinai belonged to St. Katherine's monastery since the time of Justinian (early 6th ctry.). This was reconfirmed by the Ottoman Sultans, but in Burckhardt's time this claim had become meaningless, since the government had no control over the Bedouin tribes of Sinai.

[11] 32 ancient threshing floors were recorded on the eastern side of 'Uvda Valley, built in the 4th millennium B.C. and reused in later periods. This cluster of threshing floors is the largest and oldest known to date in the Near East. Excavation in two of them yielded the earliest indications for the use of the animal-drawn threshing sledge, and the oldest stone plow tip (see Avner *et al.* 2003).

third or half of the yield, and they also sold grain in 'Aqaba and even in al'Arish. Another portion of the yield was distributed in the tent camps throughout their territory, and the remainder was stored in silos within the valley. The stored grain was kept to support the tribe for 2-3 years without flood, and the *Haywat* claimed that more than 4000 people subsisted on the yields.

Although there is some doubt as to the accuracy of these figures, there are several indications that they are not far from reality.[12] Today, four main clusters of silos (*mantara*) excavated in the ground are known in the area (Fig. 30), each cluster consisting of 60-110 silos. Together with some small clusters and individual silos, this accounts for at least 400 excavated silos in 'Uvda. According to the elders, all silos belonged to the *Haywat*, and each was excavated to hold three tons of clean grain (ten *quntar* in their measuring system). In addition, they had another *mantara,* in Nahal Girzi (*Wadi Rabiyeh*), 10 km north-west of 'Uvda, with about 60 shallow silos consisting of cleared circles, 3 m in diameter (Fig. 31). Here they laid first a layer of straw, then built a pile of grain, which they covered with straw and beaten earth (Fig. 32).

Calculations show that the total capacity of the excavated silos in 'Uvda Valley alone was over 1200 tons of grain. If a bedouin family of five consumed approximately 500 kg annually,[13] then the amount of grain stored after a successful season could have supported at least 2400 families. But, since this yield was to be used over a 2-3 year period, as well as for sowing, we reach an estimation of 800 families or 4000 people. This figure is four times higher than the known tribe's population before 1948 (see above).[14]

Fig. 30. Northern 'Uvda Valley, part of mantara, *a cluster of Bedouin silos excavated into the ground; an aerial photo.*

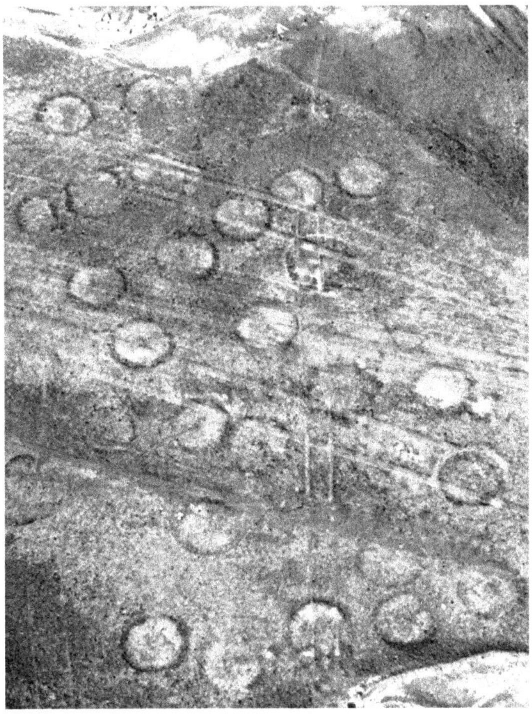

Fig. 31. Nahal Girzi, part of a mantara *with bases for grain heaps; an aerial photo.*

It seems therefore that 'Uvda Valley was indeed an important food source for the *Haywat*. According to the testimony of Kibbutz Yotvata members, the *Haywat* continued to penetrate the Sinai border from the Kuntila area as late as 1957, in attempts to cultivate the valley. During our visit to the valley with the elders (when the

[12] The traditional Bedouin agriculture in the Be'er Sheva area reached yields of 200-1000 kg wheat per hectare and slightly higher yields of barley (Ben David 1988:46-47). In this area, the agriculture was based on direct rain irrigation, which is usually less productive than flood irrigation. In the traditional agriculture of the Petra area, which generally enjoys good amounts of rain, the wheat yields are similar to those of the Be'er Sheva area, while barley yields are between 200-2000 kg per hectare (Russell 1995:697).

[13] An average annual grain consumption of a present-day seven-person rural family in the Levant is approximately 700 kg, and 500-700 kg for a seven-person Arab family is reported from the Samaria area (Dar 1982:327). Consumption of 800 kg of grains for a six person Bedouin family, is documented in Be'er Sheva' area, when grains were subsidized and therefore consumption was higher (Ben David 1982:180).

In several studies higher values of annual grain consumption were adopted, 140-230 kg per person (*e.g.* Broshi 1986:42, with refs.; Zorn 1994:43, with references.; Gregoire 1999:30-31). In my opinion, these figures are too high, for the following reasons: 1. They usually address adult consumption, while more realistic calculations should take the average between adults and children. 2. Baked bread contains up to 50% water. 200 kg of grain equates *ca.* 400 kg of bread, *i.e.* 571 present-day loaves per person per year. This means 1.5 loaves per day, or 2600 kcal., almost the total daily calory consumption for a working person. Even in a society that subsists on a limited food basket, there is no justification for such a high bread consumption, certainly not for a society which subsists on a combined economy, of herding, fishing and agriculture, as in the case of the *Haywat*.

[14] In order to fully appreciate the reliance of the tribe on agriculture, the nature of rainfall in the Southern Negev and Eastern Sinai must be considered. Concentrated rains that create floods usually fall in the autumn and spring, brought by the "Red Sea depressions". Winter rains are usually moderate, brought by the Mediterranean depressions, and do not create floods. The concentrated rains are restricted to geographic "cells", which means that in any given year one part of the tribe's territory may have flood and the others do not. During years of no flood in 'Uvda Valley, cultivated fields in the other parts of the tribe's territory could still have compensated for the want. For the nature of desert rains see Shanan *et al.* 1967; Finkel & Finkel 1979; Sharon 1979.

Fig. 32. Grain heaps near Be'er Sheva', covered by straw and beaten earth (Marx 1967: Pl. 3).

Fig. 33. Some statues of Salem.

'Uvda air-base was under construction), they repeatedly lamented the loss of "*Wadi 'Igfi, the best soil of the Haywat*".

Bedouin Sculpture

Artistic sculpturing is very rare in Arab societies, due to a virtual Islamic prohibition.[15] Some forms of art, which represent a spiritual domain, do exist in the bedouin society, in weaving, embroidery, basketry (see Figs 17, 22, 23), as well as poetry, music (Fig. 20) and dancing. Commonly, these types of art are quite formal, but there are exceptions.

When Israelis started to visit Sinai in 1967, they would occasionally meet bedouin selling small stone statutes of animals. One of them was Salem Hamid of the *Haywat* who manufactured and sold millstones, as well as small statutes, despite the elders' ban. He was "discovered" by Alfonso Nussbaumer who suggested that he sell statues near a tourist attraction. Salem settled in Wadi Umm Sidra (the "Inscription Canyon") west of Eilat, and later moved to Wadi Merakh, by the Red Sea coast. With time his statues grew larger, up to 70 cm (Fig. 33); he sold many hundreds to tourists, and his statues even reached several exhibitions (Goren and Eben 1990). In 1977 they were exhibited in New York, and the *New York Times* published an interview with him (by William Parrell). Salem's statues are somewhat "naive" in style (obviously he never studied in any art academy), but they are rich in imagination and symbolism. Today, Salem lives in Wadi Twaibeh, 3 km south of the Taba border passage, and he is still sculpturing. Salem is certainly a phenomenon, since no other tribe in Sinai produced any real sculpture. This, and the fact that the *Haywat* elders eventually removed the ban from him, may mean something about this small tribe.

Conclusion

Despite the initial low esteem for the *Haywat*, a close look at the tribe reveals a society with traditions, a social order reflected in their legal system, a firm material culture with artisans skilled in several crafts, and even successful agriculture. Their spiritual world is also notable, as seen through their crafts, poetry, sculpture and other forms of expression. Since they lived in a small territory, on a mixed economy, they cannot be considered nomads, and certainly, they do leave remains.

If this is true for a small tribe, it is obviously true for the bedouin society in general. More so, this argument should be applied to ancient desert societies as well, who also suffer from scholarly misconceptions.[16]

Acknowledgments

I am grateful to A. Nussbaumer for allowing me to photograph the bedouin artifacts from his collection, and helping me reading German texts. I am also thankful to M. Shemtov, G. Levinstein and E. Saragusti for sharing their findings in the field, and to M. Shemtov and R. Avner for proofreading the draft. Thanks to O. Goren and A. Nussbaumer for proving the bedouin names of artifacts. Photographs are mine, unless otherwise indicated.

Bibliography

Abu-Rabia', A.
1994 *The Negev Bedouin and Livestock Rearing.* Oxford: Berg.

[15] The Qur'an often quotes Muhammad condemning the worship of idols, or fighting against idolaters (*e.g.* 2:22, 220; 4:51; 7:36; 9;115; 22:26-37; 28:64; 35:38), but there is no prohibition on sculpturing in general. This appeared later in the *hadith* and other types of literature (see Wensinck 1997, with references).

[16] In addition to the introduction, above, references to scholars' views of past desert societies, and criticisms, see Avner *et al.* 1994; Avner 2002b and in press.

Al 'Aref, 'A.
1937 *The Bedouin Tribes of the Be'er Sheva' District.* Translated from Arabic to Hebrew by M. Kapeliuk. Tel Aviv: M. Kapeliuk.

Avner, U.
1990 "Ancient Agricultural Settlement and Religion in the Uvda Valley in Southern Israel". *Biblical Archaeologist* 53: 125-141.
1998 "Settlement, Agriculture and Paleoclimate in 'Uvda Valley, Southern Negev Desert, 6th-3rd Millennia B.C." in *Water, Environment and Society in Times of Climate Change*, eds. A. Issar and N. Brown. Dordrecht: Kluwer Academic Publishers. pp. 147-202.
2002a "Ancient Water Management in the Southern Negev". *ARAM Periodical* 13-14:389-407.
2002b Studies in the Material and Spiritual culture of the Negev and Sinai Population,i During the 6th-3rd millennia B.C. Unpublished Ph.D. dissertation. Jerusalem: Hebrew University.
In press "Settlement Pattern in the 'Araba Valley and the Adjacent Desert Areas, a View from the Eilat Region" in *Crossing the Rift: Resources, Routes, Settlement pattern and interaction in the Wadi Arabah,* eds. P. Bienkowski and K. Galor.

Avner, U., Carmi, I. and Segal, D.
1994 "Neolithic to Bronze Age Settlement of the Negev and Sinai in Light of Radiocarbon Dating, A View From the Southern Region" in Late Quaternary and Paleoclimates of the Eastern Mediterranean, eds. O. Bar-Yosef, and R.S. Kra. *Radiocarbon.* Tucson: University of Arizona. pp. 265-300.

Avner, U., Anderson, P.C., Mai, B., Chabot, J. and Cummings, L.
2003 "Ancient Threshing Floors, Threshing Tools and Plant Remains in 'Uvda Valley, Southern Negev Desert, Israel. A Preliminary Report" in *Le traitement des Récoltes: un regard sur la diversité, du Néolithique au présent XXIIIe rencontres internationales d'archéologie et d'histoire d'Antibes*, eds. P.C. Anderson, L.S. Cummings, T.K. Schippers and B. Simonel. Antibes: Editions APDCA. pp. 455-75.

Avni, G.
1992 "Survey of Deserted Bedouin Campsites in the Negev Highlands and its Implications for Archaeological Research" in *Pastoralism in the Levant*, eds. O. Bar Yosef and A. Khazanov. Madison, Wis.: Prehistory Press. pp. 241-54.

Banning, E. B. and Köller-Rollefson. I.
1992 "Ethnographic Lessons for the Pastoral Past: camp Locations and Material Remains Near Beidha, Southern Jordan" in *Pastoralism in the Levant*, eds. O. Bar Yosef and A. Khazanov. Madison, Wis.: Prehistory Press. pp. 181-204.

Bartlett W. H.
1849 *Forty Days in the Desert on the Track of the Israelites.* London: Virtue.

Beit -Arieh, I.
1984 "New Evidence of the Relations Between Canaan and Egypt During the Proto-Dynastic Period". *Israel Exploration Journal* 34:20-23.

Ben David, J.
1982 Stages in the Sedentarization of the Negev Bedouin, a Transition from Farmer Semi Nomadic to Settled Population. Unpublished Ph.D. dissertation. Jerusalem: Hebrew University (Hebrew).
1988 *Agricultural Settlements for the Beduin Population – Policy Proposal.* Jerusalem: The Institute for Israel Studies (Hebrew).

Broshi M.
1986 "The diet of Palestine in the Roman period. Introductory notes". *Israel Museum Journal* 5:41-56.

Burckhardt, J. L.
1822 *Travels in Syria and the Holy Land.* London: Murray.

Cribb, R.
1991 *Nomads in Archaeology.* Cambridge: Cambridge University.

Dar, S.
1982 *The Settlement Extent in Western Sameria During the Roman and Byzantine Periods.* Tel Aviv: Society for Protection of Nature (Hebrew).

Eldar, I.; Nir, Y.; and Nahlieli, D.
1992 "The Bedouin and Their Campsites in the Dimona Region of the Negev: Comparative Model for the Study of Ancient Desert Settlements" in *Pastoralism in the Levant*, eds. O. Bar Yosef and A. Khazanov. Madison, Wis.: Prehistory Press. pp. 205-17.

Finkel, H., and Finkel, M.
1979 "Surface Flow in the Arava" in *The Land of the Negev*, eds. A. Shmueli and J. Grados. Tel Aviv: Ministry of Defense (Hebrew). pp. 125-39.

Finkelstein, I.
1995 *Living on the Fringe.* Sheffield: Sheffield Academic Press.

Finkelstein, I., and Perevolotsky, A.
1990 "Processes of Sedentarization and Nomadization in the History of Sinai and the Negev". *Bulletin of the American Schools of Oriental Research* 279: 67-88.

Frank, F.
1934 "Aus der 'Araba". *Zeitschrift des Deutschen Palästina Vereins* 57: 191-280.

Glueck, N.
1935 "Explorations in Eastern Palestine II". *Annual of the American Schools of Oriental Research* 15: 1-202.
1968 *Rivers in the Desert* (revised edition). New York: Norton Library.
1970 *The Other Side of Jordan.* Cambridge, MA: American Schools of Oriental Research.

Goren-Inbar, N.
1993 "Ethnoarchaeology: The Southern Sinai Bedouin as a Case Study" in *Biblical Archaeology Today, 1990, Proceedings of the Second International Congress on Biblical Archaeology*, eds. A. Biran, and J. Aviram. Jerusalem: Israel Exploration Society. pp. 417-19.

Goren, O., and Eben, D.
1990 "Statues from the Desert". *Studio* 11: 28-30 (Hebrew).

Gregoire, J.P.
1999 "The Grain-Grinding-Households (e2-HAR.HAR) of Southern Mesopotamia at the End of the 3rd Millennium Before the Common Era". *Bulletin of the Anglo-Israel Archaeological Society* 17: 7-38.
Haiman, M.
1992a "Sedentarism and Pastoralism" in *Pastoralism in the Levant*, eds. O. Bar Yosef and A. Khazanov. Madison, Wis.: Prehistory Press. pp. 93-104.
1992b Early Settlement Patterns of the Negev Highlands. Analsis of the Findings of the Negev Emergency Survey 1979-1989. Unpublished Ph.D. dissertation. Jerusalem: Hebrew University.
Jarvis C. S.
1931 *Yesterday and Today in Sinai*. London: Blackwood.
1936 *Three Deserts*. London: Murray.
1938 *Desert and Delta*. London: Murray.
Jennings-Bramley, W.E.
1905 "The Bedouin of the Sinaitic Peninsula". *Palestine Exploration Quarterly* 38: 126, 211.
1906 "The Bedouin of the Sinaitic Peninsula". *Palestine Exploration Quarterly* 39: 23,103,197,250.
1907 "The Bedouin of the Sinaitic Peninsula". *Palestine Exploration Quarterly* 40: 22,131, 279.
1908 "The Bedouin of the Sinaitic Peninsula". *Palestine Exploration Quarterly* 41: 30,112.
1909 "The Bedouin of the Sinaitic Peninsula". *Palestine Exploration Quarterly* 42: 253.
1910 "The Bedouin of the Sinaitic Peninsula". *Palestine Exploration Quarterly* 43: 140.
1911 "The Bedouin of the Sinaitic Peninsula". *Palestine Exploration Quarterly* 44: 34,172.
1912 "The Bedouin of the Sinaitic Peninsula". *Palestine Exploration Quarterly* 45: 13,62.
1914 "The Bedouin of the Sinaitic Peninsula". *Palestine Exploration Quarterly* 47: 9.
Khazanov, A.M., and Bar-Yosef, O.
1993 "Anthropological Aspects of Recent Archaeological Research in Pastoralism" in *Biblical Archaeology Today, 1990, Proceedings of the Second International Congress on Biblical Archaeology*, eds. A. Biran, and J. Aviram. Jerusalem: Israel Exploration Society. pp. 459-64.
Marx, E.
1967 *Bedouin of the Negev*. Manchester: Manchester University.
1992 "Are There Pastoral Nomads in the Middle East?" in *Pastoralism in the Levant*, eds. O. Bar Yosef and A. Khazanov. Madison, Wis.: Prehistory Press. pp. 255-60.
1995 "Are there pastoral nomads in the Arab Middle East?" in *The Anthropology of Tribal and Peasant Pastoral Societies*, eds. U. Fabietti and Ph. C. Salzman. Pavia: Collegio Ghislieri and Como: Ibis. pp. 71-85.
Murray, G. W.
1935 *Sons of Ishmael: A Study of the Egyptian Bedouin*. London: Routledge & Sons.
Musil, A.
1907-8 *Arabia Petraea* Vols I-III. Wien: Alfred Holder.
1926 *The Northern Hegaz*. New York: American Geographical Society.
Robinson, E. and E. Smith
1856 *Biblical Researches in Palestine and the Adjacent Regions I-IV*. London: Murray.
Rüppell, E.
1829 *Reisen in Nubien, Kordofan und dem Peträischen Arabia*. Frankfurt am Main: Friedrich Williams.
Russell, K. W.
1995 "Traditional Bedouin Agriculture at Petra: Ethnoarchaeological Insights into the Evolution of Food Production". *Studies in the History and Archaeology of Jordan V*. Amman: Department of Antiquities: 693-705.
Saidel B.
2001 "Abandoned Tent Camps in Southern Jordan". *Near Eastern Archaeology* 64/3: 54-61.
Shanan L.; Evenari, M.; and Tadmor, N.
1967 "Rainfall Pattern in the Negev Desert". *Israel Exploration Journal* 17: 163-184.
Sharon, D.
1979 "Local Rain Storms in the Arava Valley" in *The Land of the Negev*, eds. A. Shmueli and J. Grados. Tel Aviv: Ministry of Defense (Hebrew). pp. 119-24.
Shkolnik, A.
1977 "Physiological Adaptation of Mammals to Life in the Desert" in *The Desert, Past, Present, Future*, ed. E. Sohar. Tel Aviv: Reshafim. (Hebrew). pp. 100-12
Stewart, F. H.
1977 *Fundamentals of age-group systems*. New York: Academic Press.
1986 *Bedouin boundaries in Central Sinai and the Southern Negev: a document from the AHaywat tribe*. Wiesbaden: Harrassowitz.
1988 *Texts in Sinai Bedouin law*. Wiesbaden: Harrassowitz.
1994 *Honor*. Chicago: University of Chicago.
2003 *The contract with surety in Bedouin customary law*. UCLA Journal of Islamic and Near Eastern Law 2: 163-280.
Von Oppenheim, M.
1939-1952 *Die Beduinen I-III*. Leipzig: Harrassowitz.
Wensinck, A. J.
1997 "Sura" in *The encyclopedia of Islam* IX. Leiden: Brill. pp 889-92.
Winter, M.
1987 "The Sinai Peninsula Under the ottoman Rule" in *Sinai* Vol II, eds G. Gvirtzman; A. Shmueli; Y.Gradus; I. Beit-Arieh; and M. Har-El. Tel Aviv: Ministry of Defense. (Hebrew). pp. 797-806.
Woolley, C.L., and Lawrence, T.E.
1914-15 *The Wilderness of Zin*. London: Palestine Exploration Fund Annual III.
Zorn, J.R.
1994 "Estimating the Population Size of Ancient Settlements: Methods, Problems, Solutions, and a Case Study". *Bulletin of the American Schools of Oriental Resaerch* 295:31-48.

RUNOFF TERRACES IN THE NEGEV HIGHLANDS DURING THE IRON AGE: NOMADS SETTLING DOWN OR FARMERS LIVING IN THE DESERT?

Hendrik J. Bruins

Abstract

The Iron Age desert village of Horvat Haluqim in the Central Negev Highlands is an example of many similar Iron Age sites in the region, which often include a kind of fortress that is roughly elliptical in outline, having casemate rooms around a central courtyard, ranging in size from *ca* 20 to 60 m diameter. Were these Iron Age settlements established by societies with a pastoral nomadic background or with a distinct agricultural tradition? The very existence of dry stone walls across the wadi to arrest the runoff flows to increase soil moisture is a prime characteristic for an agricultural society. Geoarchaeological excavations in a terraced field in the eastern wadi of Horvat Haluqim revealed a buried anthropogenic soil layer across the entire width of the field. The first radiocarbon dates from the upper part of this layer indeed confirmed Iron Age agricultural activity as early as possibly the 11th century BCE. The latter date is based on a bone from a sheep or goat. Micromorphology investigations of the anthropogenic layer revealed iron oxide nodules, which are clear evidence of sustained runoff–farming practices in the past. The occurrence of tiny charcoal and bone particles at the microscopic level is strong evidence for the use of home refuse as a fertiliser. Moreover, the presence of high concentrations of phytoliths indicates the past application of animal manure on the fields, also to increase soil fertility. Such manuring practices, characteristic for agricultural societies, are not carried out by the modern Azazmeh Bedouin in the area, even though they engage in marginal agriculture. Therefore, all geoarchaeological data linked to comparative anthropology suggest that the Iron Age villages in the Negev Highlands were established by people with a thorough agricultural tradition of mixed farming rather than by nomadic pastoralists.

Introduction

The central Negev highlands and adjacent area of north-eastern Sinai contain many archaeological remains belonging to the Iron Age. These settlements often include elliptical or irregular shaped fortresses and smaller dwelling units (Cohen 1980, 1986, 1993; Haiman 1994). Different scholarly opinions exist concerning the date and function of these settlements, particularly the fortresses. Their archaeological age assessment ranges from the 13th or 12th century (Rothenberg 1967, 1972, 1988, 1999) to the 7th century BCE (Aharoni *et al.*, 1960; Aharoni 1967). Later Aharoni (1978) changed his opinion and favoured the 11th century BCE, an age also preferred by Herzog (1983). The lack of unambiguous archaeological dating criteria is underlined by Meshel (1979), who considered both the 11th and 10th century as options (Meshel and Goren (1992). Finkelstein (1984, 1986, 1995) favoured the 11th and early 10th century BCE for the age of the fortresses. A 10th century BCE date is strongly advocated by Cohen (1980, 1986, 1993) and Haiman (1994).

Concerning the origin and character of these sites, many scholars suggested associations with various portions of the Biblical text, ranging from the possible settlement of the tribe of Simeon (Herzog 1983) to the United Monarchy (Cohen 1980, 1986, 1993; Haiman 1994). An Amalekite origin was proposed by Rothenberg (1967, 1972, 1988), also in relation to the biblical text. Finkelstein (1984, 1986, 1995) interpreted the settlements as sedentarization of local nomadic desert tribes that became involved in international trade.

The excavation of building remains and the study of architectural elements and ceramic styles understandably form the traditional focus of archaeology, *i.e.* the material remains of ancient societies. However, the landscape and the soil were the focus of the ancient societies themselves in terms of land-use, food production and water resources. Such relationships between environment and archaeology (Butzer, 1971) need to be investigated as well. Terraced wadis in the central Negev constitute the most prominent mark of landscape changes brought about by people in the past in order to enhance local resource management with the utilization of runoff water for improved plant growth in the desert. The nature of these plants may range from cereal food crops to fruit trees (Evenari *et al.*, 1971). Geoarchaeological excavations in terraced fields in the Negev Highlands were initiated by Bruins at Horvat Haluqim and Nahal Mitnan (1986, 1990a, 1990b).

This paper concisely addresses the unique "archive" of past human activities as preserved in the ancient anthropogenic soil of the terraced fields. The character of some of the findings of these geoarchaeological investigations is related to the question whether the Iron Age settlements in the Negev Highlands belonged to

Dryland Zone	P/PET Ratio	Interannual Rainfall Variability	Possible Rainfed Land-use
Hyper-arid	< 0.05	~ 100%	Limited livestock grazing, mainly by camels and goats Runoff irrigated agriculture rare
Arid	0.05 - 0.20	50-100%	Classical livestock grazing zone (less suited for cattle) Runoff irrigated agriculture feasible
Semi-arid	0.20 - 0.50	25-50%	Rainfed Agriculture (drought risk serious) Livestock grazing (all species)
Dry Sub-humid	0.50 - 0.65	< 25%	Rainfed Agriculture (drought risk significant) Livestock grazing (all species)

Table 1. The dryland zones and their characteristics. Source: Unesco (1979), Middleton and Thomas (1997), Bruins and Lithwick (1998).

pastoral or agricultural societies. Since chronology forms an important part in archaeological theory building (Bruins *et al.*, 2003a), the issue of radiocarbon dating (Bruins and Van der Plicht, 2004, 2005) is also briefly addressed.

The Negev Highlands, Aridity and Land-Use

Modern classifications of drylands generally distinguish four different climatic zones (Table 1): hyper-arid, arid, semi-arid, and sub-humid. The P/PET index forms the basis for the classification, in which P is the average annual precipitation and PET the average annual potential evapotranspiration (Unesco, 1979; Middleton and Thomas, 1997; Bruins and Berliner, 1998). The value for PET is calculated according to the Thornthwaite approach (Thornthwaite and Mather 1957), as advocated by Middleton and Thomas (1997), recently employed for the first time in dryland classification studies in Israel (Kharel and Bruins 2004). The true arid frontier within the drylands in terms of human ecology and settlement is the boundary between the arid and semi-arid zone. The P/PET value for this boundary is 0.20 (Table 1).

The semi-arid and sub-humid zones are suitable for rainfed agriculture, albeit with significant risks of crop failure in drought years (Bruins and Lithwick, 1998). The climate in the central Negev Highlands is arid to hyper-arid (Kharel and Bruins 2004), *i.e.* too dry for common rainfed agriculture, but suited for extensive pastoralism. However, a rather unique form of agriculture developed in the central Negev Highlands: runoff farming, also termed rainwater harvesting agriculture (Bruins 1986, 2003; Bruins et al 1986). The natural division of rocky hills and small valleys with rather thick loessial soils enabled irrigation with runoff water derived from local rainfall. Hundreds of ephemeral stream channels (wadis) were terraced in ancient times with check-dams made of dry stone walls to arrest the runoff flows and increase the amount of soil moisture in the subsoil (Kedar 1957; Evenari *et al.* 1971; Bruins 1986, 1990a, 1990b, 2003).

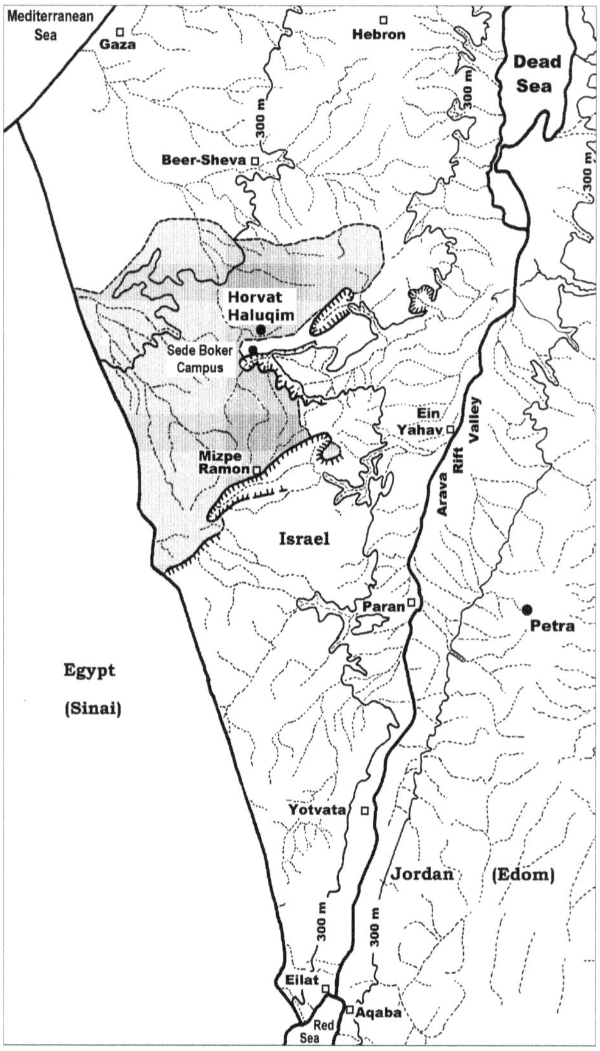

Fig. 1. The Central Negev Highlands (shaded) are characterized by many remains of ancient runoff agriculture from different periods including the Iron Age. The contour line of 300 m gives a rough indication of the topography.

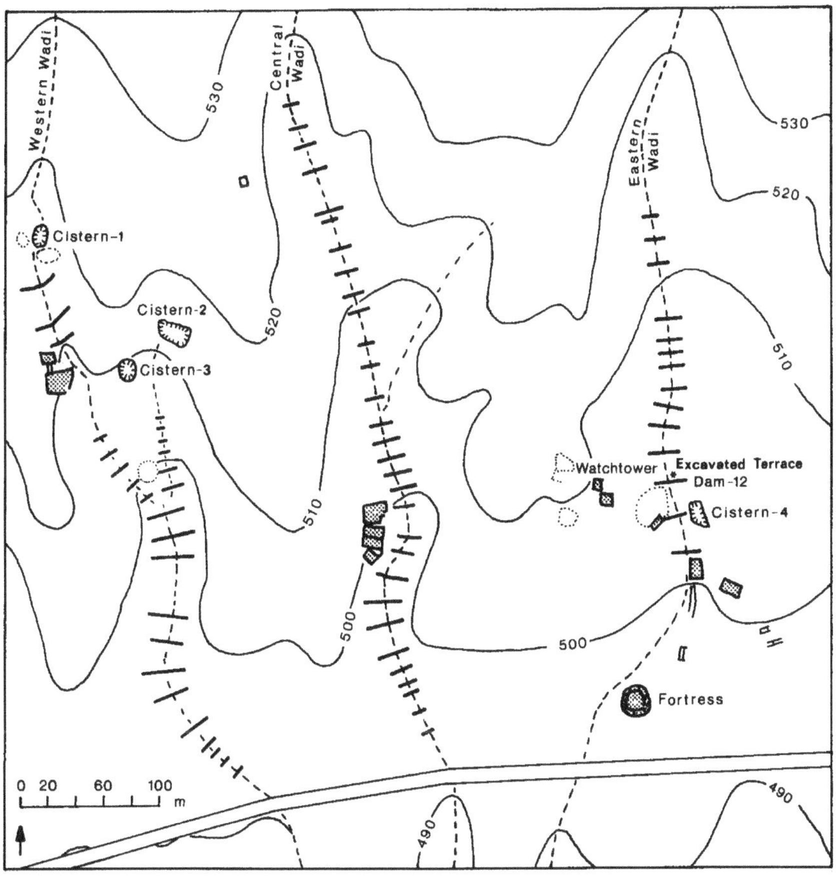

Fig. 2. The site of Horvat Haluqim with the three terraced wadis (source: Cohen, 1976 and Bruins, 1986)

Geoarchaeological excavations at the Iron Age Site of Horvat Haluqim

The large number of Iron Age settlements in the central Negev Highlands is underlined by the classification of 350 sites by Haiman (1994) in the western part of the region. Horvat Haluqim is situated in the eastern part of the area on the Haluqim Anticline about 5 km north of the Sede Boker Campus (Fig. 1). Cohen (1976) carried out archaeological excavations at Horvat Haluqim, a rural desert village, located along three small parallel wadis that descend on the south-eastern slopes of the Haluqim Anticline (Fig. 2). The village consisted of 25 structures, including an oval fortress (23 m diameter), seven houses of the four-room type, a rectangular public building (14 x 9 m) and four cisterns (Cohen 1993).

Indeed the largest building at Horvat Haluqim is the oval fortress. There are many such Iron Age fortresses in the central Negev highlands (Cohen 1980; Haiman 1994; Cohen and Cohen-Amin, 2004); a total of 58 were classified by Haiman (1994) in surveys of the western part of the region. Though the word "fortress" primarily suggests a military or policing function, while other roles cannot be excluded, the terms is maintained in this paper for such buildings as currently the most common descriptive word.

The fortress at Horvat Haluqim (Cohen, 1976) has an oval ground plan with diameter of 23 x 21 m. Double casemate walls, subdivided in seven rooms, surround a central courtyard. The gate was located at the south-east side of the fortress. Only the foundations of the walls were preserved, built on bedrock composed of Turonian limestone. Roughly hewn blocks of local limestone were used in the building. Handmade cooking-pots, so-called Negbite ware, were the most common type of vessel found in the fortress. Cohen (1976) suggested a 10th century date for the fortress in relation to the reign of Solomon, based on wheelmade pottery. However the Negbite type of pottery has a long history, going back into the Late Bronze Age (Rothenberg, 1967, 1972, 1988, 1999). Radiocarbon dates of destruction layers in relation to two other oval fortresses, at Tell el-Qudeirat in north-eastern Sinai and near Nahal Ha'Elah in the central Negev, yielded dates that seem distinctly older than the Solomonic period (Bruins and Van der Plicht, 2005).

The ancient agricultural fields were surveyed by Bruins (1986) to measure their respective sizes. More than 70 stone terrace walls were found in the three wadis of Horvat Haluqim. Excavations in five different parts of the 12th terraced field of the eastern wadi revealed a darker soil layer beginning at a depth of about 45-50 cm below the present surface. This anthropogenic agricultural soil layer has a remarkable thickness of about 75 cm. It yielded some Iron Age pottery sherds, as well as small pieces of charcoal and small animal bones, which were used for radiocarbon dating at the University of Groningen (Bruins and Van der Plicht 2004).

Radiocarbon Dating

A small bone from a sheep or goat, found inside the anthropogenic soil layer, gave an AMS date of 2860 ± 40 BP (GrA-14398). A bone date can be regarded as short-lived, because sheep or goats in pastoral systems usually do not live longer than a few years. The 1σ calibrated age is 1124-1121 (0.9%), 1111-1099 (4.7%), 1080-1061 (7.5%), 1052-971 (45.7%), 958-938 (9.4%) years BCE, and the 2σ range is 1189-1179 (1.9%), 1154-1142 (1.5%), 1129-915 (92.0%) years BCE. The most likely period within all these options is 1052-971 BCE with a relative probability of 45.7% in the 1σ range (Bruins and Van der Plicht 2004). This would indicate the practising of ancient runoff agriculture at Horvat Haluqim, including manuring of the studied terraced field, prior to the Solomonic period. In fact the above result is backed by two radiocarbon dates of destruction levels from the oval fortress at Tell el-Qudeirat and at Nahal Ha'Elah, which favour Iron Age I rather than Iron Age II (Bruins and Van der Plicht, 2005).

A small charcoal fleck, also from the anthropogenic soil layer, yielded a date of 2590 ± 60 BP (GrA-12448) in the later Iron Age. The 1σ calibrated date is 828-759 (41.2%), 682-666 (6.6%), 634-591 (14.1%), 578-557 (6.4%) and for 2σ the result is 896-875 (1.8%) and 842-519 (93.6%) years BCE. The highest relative probability in the 1σ range is for the period 828-759 BCE. Therefore, human activity at the agricultural terrace *also* occurred during the later Iron Age. The old-wood effect of the tiny charcoal fleck is probably small. Trees do not grow in most of the region, while little fires for cooking purposes can be made from local shrubs, as the Bedouin do today. Such organic material from shrubs is likely to be rather short-lived.

Micromorphology

More than 10 undisturbed soil samples were taken from this anthropogenic terrace layer for microscope studies. Firm natural soil peds (aggregates) that do not fall apart are suitable; their size should be 5 cm or more to enable micro-stratigraphic analysis. However, soil samples were generally extricated with a special technique into small boxes to preserve their natural coherence, as the anthropogenic soil is usually without a clear structure and therefore without peds. The vertical orientation of the samples was marked in the field. Afterwards, the samples were immersed with a liquid resin under vacuum in the Laboratory for Soil Science and Geology of Wageningen University (The Netherlands) to impregnate the soil. The resin penetrates all voids and replaces the air, while gradually becoming very hard (see for example FitzPatrick 1970). Thus thin sections can be made from undisturbed soil samples with marked field orientation for studies with a petrographic microscope.

The micromorphological study revealed the presence of ferric nodules in the anthropogenic soil layer of the terraced field, having a diameter of 0.25 to 0.35 mm (Bruins 1986; Bruins and Jongmans 2004). Iron is only mobile in the soil under a wet and reduced chemical environment. This is not typical for a desert environment. However, the building in ancient times of the 12th dam in the eastern wadi of Horvat Haluqim blocked part of the runoff flows and caused temporary flooded conditions on this field. Thus ferric nodules could gradually form in the course of time, as long as people maintained this check-dam at a higher level than the surface of the terraced field. Therefore, the ferric nodules are first-class evidence of the sustained practice in the past of runoff agriculture.

Microscopic research revealed many tiny charcoal particles and some bone remnants often smaller than one mm in size. In fact the presence of silt size charcoal particles dispersed in the anthropogenic soil layer causes the comparatively dark colour. The presence of both charcoal and bone remnants indicated that the local inhabitants used home refuse to manure the soil in order to increase the soil fertility, which is normally low in a desert environment.

Groningen AMS Lab sample no.	Organic material	Date in conventional ^{14}C years BP	1σ calibrated date years BCE	2σ calibrated date years BCE
GrA-14398	Bone of goat or sheep	2860 ± 40	1124-1121 (0.9%) 1111-1099 (4.7%) 1080-1061 (7.5%) 1052-971 (45.7%) 958-938 (9.4%)	1189-1179 (1.9%), 1154-1142 (1.5%), 1129-915 (92.0%)
GrA-12448	Charcoal fleck	2590 ± 60	828-759 (41.2%), 682-666 (6.6%), 634-591 (14.1%), 578-557 (6.4%)	896-875 (1.8%) 842-519 (93.6%)

Table 2. Radiocarbon dates from Terrace 12 at Horvat Haluqim

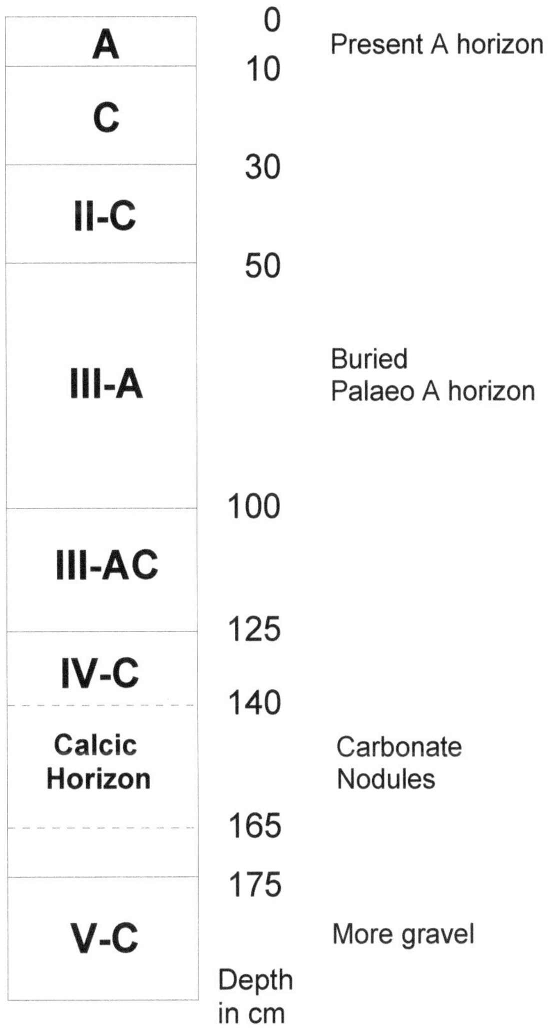

Fig. 3. Geoarchaeological soil profile stratigraphy in Pit 1 of the 12th terraced field in the eastern terraced wadi at Horvat Haluqim.

Spherulites
Sheep and goats, as well as other ruminants, produce in their intestines very small rounded crystalline calcite particles in the fine silt fraction, *ca* 5-20 micron in size, termed spherulites. Subsequently, these spherulites are excreted with the faeces and appear in the dung of these animals. Spherulites, as mineral particles, can remain in the soil for long periods of time after the dung has disintegrated and disappeared (Canti, 1997, 1998, 1999). Hence the qualitative and quantitative presence of spherulites in soils and other geoarchaeological contexts may convey significant information concerning ancient land-use, pastoralism or mixed farming.

Four soil samples from the agricultural terraces at Horvat Haluqim were investigated by Albert, who developed a quantitative approach to study phytoliths and spherulites (Albert and Weiner, 2001). Though this information will be published elsewhere in more detail, it appeared that the upper part of the comparatively dark anthropogenic soil layer contains 12 times more spherulites than the present modern soil surface. This is strong evidence that animal manure from sheep and/or goats was used in the Iron Age to fertilize the soil. Such a practice of using the accumulating animal dung in the corrals to manure the agricultural fields is not done by the modern Bedouin living in the area, largely belonging to the Azazmeh tribe. Though these Bedouin engage in marginal agriculture, usually practised on the remains of ancient runoff fields, they do not manure the soil, even though manure is available to them from their herds in the corrals! This underlines the fact that the Azazmeh Bedouin have a nomadic pastoral tradition and not an agricultural tradition.

Discussion and conclusions

It was suggested by Finkelstein (1986, 1995) that the Iron Age fortresses in the central Negev Highlands lacked a military function but were established by local nomads, as they settled down to a sedentary way of living. Finkelstein (1986: 48) tried to make a suggested analogy with the Bedouin in the region in modern times, as he pointed to a photograph from Dalman (1939) in which Bedouin tents are placed together in a kind of rectangular alignment that, in his view, resembled the layout of the fortresses: “a chain of rooms (like tents) ringing an oval enclosure for the herds”.

There are several fundamental problems with this viewpoint. First, Bedouin tents are usually *not* positioned close to each other, while having a *preferred north-south orientation.* Based at the Sede Boker Campus, I have never seen such a close alignment of Bedouin tents during the 1980s and 1990s in my extensive travelling in the region, including the Negev, Sinai and south-western Jordan. The photo from Dalman (1939) seems an exception in comparison to the later decades of the 20th century. However, such an alignment of Bedouin tents around an oval space was apparently more common in the region during the 19th century (Eveline van der Steen, personal communication, 2005).

Second, the occasional alignment of tents in a rectangular or oval way around a central courtyard does not make such examples ‘prototypes’ for stone buildings with casemate walls around a central courtyard. The modern Bedouin in their recent process of sedentarization and spontaneous settlement (Ben-David, 1982) did not build such large stone buildings as the Iron Age fortresses, neither in size nor in shape. The building plans of the Bedouin dwellings did not go beyond the needs for a nuclear or slightly extended family. Some of the many Iron Age fortresses, ranging in size from 20 to 60 meter in diameter (Cohen, 1980; Meshel and Goren, 1992), were also built on isolated hilltops. Such a position does not seem ideal for keeping the herd, if the central courtyard within the fortress was planned for such a purpose in the Iron Age. Hence the role of public security of the “fortresses” or even signifying social stratification, without a functional link to pastoralism, cannot be ruled out.

Third, the Bedouin of the Azazmeh tribe in the Negev Highlands, who during the 20th century settled down from pastoral nomadism to a sedentary way of life, keeping goats and sheep while also engaging in marginal farming, *do not apply manure* in their use of ancient terraced fields in the region (Bruins, 1986; Bruins et al., 2003b). The majority of the older Bedouin tribes in the Negev have a nomadic background without a tradition in agriculture (Kressel et al., 1991; Kressel 2003). During fieldwork and meetings with Bedouin in the region in the 1980s and 1990s, the author asked the Bedouin many times: "Do you put manure on your agricultural fields?" The answer was always negative.

The investigated terraced field at Horvat Haluqim on the other hand does exhibit unmistakable evidence of manuring in the Iron Age, as established by both radiocarbon dating and detailed micromorphological research. Who then were the Iron-Age people that established Iron Age Horvat Haluqim? They had an established tradition and experience in farming, recognizing both the need for terraced fields in order to enable runoff agriculture in the desert and the importance of applying manure to the fields to increase soil fertility and yields. The type of bones – sheep/goat – found in the ancient anthropogenic soil layer, together with the evidence from the phytoliths, strongly indicates the practice of mixed farming: a combination of pastoral livestock raising and runoff agriculture. Most of the desert region in the central Negev Highlands, *ca* 95%, is in terms of land-use only suitable for extensive grazing. Runoff agriculture can *only* be practised on about 5% of the land surface in the area. This unique form of desert agriculture is restricted to suitable wadis with deep soils and built stone walls (check-dams) to catch the runoff flows that supplied additional soil moisture for growing annual crops or fruit trees.

Bibliography

Aharoni, Y.

1967 "Forerunners of the limes: Iron Age fortresses in the Negev". *Israel Exploration Journal* 17: 1-17.

1978 *The Archaeology of the land of Israel*; translated by A. F. Rainey. Philadelphia: Westminster.

Aharoni, Y.; Evenari, M.; Shanan, L.; and Tadmor, N.H.

1960 "The ancient desert agriculture of the Negev. V. An Israelite settlement at Ramat Matred". *Israel Exploration Journal* 10:23-36, 97-111.

Albert, R.M., and Weiner, S.

2001 "Study of phytoliths in prehistoric ash layers using a quantitative approach" in *Phytoliths, Applications in Earth Science and Human History*, ed. J.D. Meunier and F. Coline. Rotterdam: Balkema. pp 251-266.

Ben-David, J.

1982 Stages in the sedentarization of the Negev Bedouin, a transition from semi-nomadic to settled population. Ph.D. thesis, The Hebrew University of Jerusalem (Hebrew with English summary).

Bruins, H.J.

1986 *Desert Environment and Agriculture in the Central Negev and Kadesh-Barnea during Historical Times.* Nijkerk: Stichting Midbar Foundation; Published Ph.D. Dissertation, The University of Wageningen.

1990a "The impact of man and climate on the Central Negev and northeastern Sinai deserts during the Late Holocene" in *Man's Role in the Shaping of the Eastern Mediterranean Landscape*, eds S. Bottema, G. Entjes-Nieborg and W. van Zeist. Rotterdam: Balkema. pp 87-99.

1990b "Ancient agricultural terraces at Nahal Mitnan". *Atiqot* 10: 127-128 (Hebrew), 22-28 (English).

2003 "Man and Landscape in the Negev Highlands: Runoff Systems". *Horizons in Geography* 57-58: 146-158 (Hebrew).

Bruins, H.J., Van der Plicht, J. and Mazar, A.

2003a "^{14}C dates from Tel Rehov: Iron-Age chronology, Pharaohs and Hebrew Kings". *Science* 300(5617): 315-318.

Bruins, H.J.; Akong'a, J.J.; Rutten, M.M.E.M.; and Kressel, G.M.

2003b *Drought Planning and Rainwater Harvesting for Arid-Zone Pastoralists: the Turkana and Maasai (Kenya) and the Negev Bedouin (Israel).* NIRP Research for Policy Series 17, The Hague, KIT Publishers, Amsterdam (ISBN 90-6832-682-1).

Bruins, H.J. and Berliner, P.R.

1998 "Bioclimatic Aridity, Climatic Variability, Drought and Desertification" in *The Arid Frontier – Interactive Management of Environment and Development*, eds H.J. Bruins and H. Lithwick. Dordrecht: Kluwer Academic Publishers. pp 97-116.

Bruins, H.J., Evenari, M. and Nessler, U.

1986 "Rainwater-harvesting agriculture for food production in arid zones: the challenge of the African famine". *Applied Geography* 6:13-32.

Bruins, H.J., and Jongmans, A.G.

2004 "Desertification or soil improvement: long-term effect of terracing and runoff irrigation on desert soils at Horvat Haluqim, Negev, Israel" in *Extended Abstracts, 12th International Meeting on Soil Micromorphology, Adana, Turkey, September 20-26, 2004,* eds S. Kapur, E. Akça, L. Montanarella, A. Ozturk, A.R. Mermut. European Commission, Joint Research Centre, EUR 21275 EN/2. pp. 13-15.

Bruins, H.J. and Lithwick, H.

1998 "Proactive Planning and Interactive Management in Arid Frontier Development" in *The Arid Frontier: Interactive Management of Environment and Development*, eds H.J. Bruins and H. Lithwick, H. Dordrecht: Kluwer Academic Publishers. pp 3-29.

Bruins, H.J. and Van der Plicht, J. 2004. "Desert settlement in the central Negev: first ^{14}C indication of rainwater-harvesting agriculture in the Iron Age" in *Radiocarbon and Archaeology: Proceedings of the 4th International Symposium,* eds T.F.G Higham, C. Bronk Ramsey, and D.C. Owen. Oxford: Oxford University School of Archaeology Monograph 62. pp 83-98.

Bruins, H.J., and Van der Plicht, J.
2005 (in press) "Desert Settlement through the Iron Age: Radiocarbon Dates from Sinai and the Negev Highlands" in T.E. Levy and T.F.G Higham (eds) *Radiocarbon Dating and the Iron Age of the Southern Levant – the Bible and Archaeology Today*. Proceedings of an International Conference held at the Oxford Centre for Hebrew and Jewish Studies, Yarnton Manor, Oxford, September 6 – 9th, 2004. London: Equinox.
Butzer, K.W.
1971 *Environment and Archeology*. Chicago: Aldine Atherton.
Canti, M.G.
1997 An investigation of microscopic calcareous spherulites from herbivore dung. *Journal of Archaeological Science* 24: 219-231.
1998 "The micromorphological identification of faecal spherulites from archaeological and modern materials". *Journal of Archaeological Science* 25: 435-444.
1999 "The production and preservation of faecal spherulites: animals, environment and taphonomy". *Journal of Archaeological Science* 26: 251-258.
Cohen, R.
1976 "Excavations at Horvat Haluqim". *Atiqot* 11: 34-50.
1980 "The Iron Age fortresses in the central Negev". *Bulletin of the American Schools of Oriental Research* 236: 61-79.
1986 The Settlement of the Central Negev in light of Archaeological and Literary Sources during the 4th-1st millennia BCE. Jerusalem: Ph.D. Dissertation, The Hebrew University of Jerusalem (in Hebrew).
1993 "Middle Bronze Age I and Iron Age II sites in the Negev highlands" in *The New Encyclopaedia of Archaeological Excavations in the Holy Land*, Volume 3, ed. E. Stern. New York: Simon & Schuster. pp 843-47.
Cohen, R., and Cohen-Amin, R.
2004 *Ancient settlement of the Negev Highlands*. Vol. II. Jerusalem: The Israel Antiquity Authority.
Dalman, G.H.
1939 *Arbeit und Sitte in Palästina. Band VI.* Gütersloh: Bertelsmann.
Evenari, M.; Shanan, L.; and Tadmor, N.
1971 *The Negev: The Challenge of a Desert*. Cambridge, MA: Harvard University (2nd edition 1982).
Finkelstein, I.
1984 "The Iron Age fortresses of the Negev highlands: Sedentarization of desert nomads". *Tel Aviv* 11: 189-209.
1986 "The Iron Age sites in the Negev highlands – Military fortresses or nomads settling down". *Biblical Archaeology Review* 12:46-53.
1995 *Living on the Fringe.* Sheffield: Sheffield Academic Press Monographs in Mediterranean Archaeology.
FitzPatrick, E.A.
1970 "A technique for the preparation of large thin sections of soils and unconsolidated materials" in *Micromorphological Techniques and Applications*, eds D.A. Osmond and P. Bullock. Harpenden: Soil Survey of England and Wales, Rothamsted Experimental Station. pp 3-13.
Haiman, M.
1994 "The Iron Age II sites of the western Negev Highlands". *Israel Exploration Journal* 44:36-61.
Herzog, Z.
1983 "Enclosed settlements in the Negeb and the wilderness of Beer-sheba". *Bulletin of the American Schools of Oriental Research* 250:41-49.
Kedar, Y.
1957 "Water and soil from the desert: some ancient agricultural achievements in the central Negev". *The Geographical Journal* 123: 179-187.
Kharel, H. and Bruins, H.J.
2004 "Drought and Desertification Hazards in Israel: Time-Series Analysis 1970-2002" in *Disasters and Society – From Hazard Assessment to Risk Reduction, Proceedings of the International Conference, University of Karlsruhe.* eds D. Malzahn and T. Plapp. Berlin: Logos. pp 91-98.
Kressel, G.M.
2003 *Let Shepherding Endure.* Albany: State University of New York.
Kressel, G.M.; Ben-David, J.; and Abu Rabi'a, A.
1991 "Changes in the land usage by the Negev Bedouin since the mid-19th century: the intra-tribal perspective". *Nomadic Peoples* 28: 28-55.
Meshel, Z.
1979 "Who built the Israelite fortresses in the Negev hills?" *Kathedra* 11: 3-29 (Hebrew).
Meshel, Z., and Goren, A.
1992 "The 'Aharoni fortress' near Quseima – Another 'Israelite fortress' in the Negev and the problem of these 'fortresses'". *Eretz-Israel* 23: 196-215.
Middleton, N., and Thomas, D.
1997 *World Atlas of Desertification.* UNEP. London: Arnold (2nd edition).
Rothenberg, B.
1967 *Zefunot Negev: Archaeology in the Negev and the Arabah.* Ramat-Gan: Masada (Hebrew).
1972 *Timna, Valley of the Biblical Copper Mines.* London: Thames and Hudson.
1988 *The Egyptian Mining Temple at Timna.* London: Institute for Archaeo-Metallurgical Studies (IAMS), University College London.
1999 "Archaeo-metallurgical researches in the southern Arabah 1959-1990. Part 2: Egyptian New Kingdom (Ramesside) to Early Islam". *Palestine Exploration Quarterly* 131:150-176.
Thornthwaite, C.W. and Mather, J.R.
1957 "Instructions and Tables for Computing Potential Evapotranspiration and the Water Balance". *Publications in Climatology*, 10(3): 205-241.
Unesco
1979 "Map of the World Distribution of Arid Regions". *Unesco, Man and the Biosphere (MAB) Technical Notes 7*, Paris.

THE TRANSFORMATION OF NABATAEAN SOCIETY: ACCULTURATION OR SELF-ORGANIZATION?

Tali Erickson-Gini

Abstract

According to historical sources and what is currently known from the archaeological record, the Nabataeans appear to have originally been a nomadic society engaged in trade on a seasonal basis. The transformation of that society in the first century BCE was dramatic and enigmatic by normal standards. What factors accounted for the transition of the Nabataeans to sedentarization in this period? At present, most researchers credit acculturation for their rapid cultural, economic and sociopolitical development. The Nabataeans were undoubtedly influenced to some degree by their contacts with neighbouring societies in the Near East, and particularly developments in the Judean kingdom in the time of Herod the Great. The relationship between economic and architectural developments in Judea and their influence on Nabataean culture is a subject that will be dealt with in this paper. However, in spite of external influences it appears that the transformation of Nabataean culture in the first century BCE was the result of the process of self-organization in which they responded to changing economic and political realities by developing a new subsistence strategy for survival.

Nomads and Traders

In the earliest historical references to the Nabataean people, dating to the last centuries of the first millennium B.C., they are described as a nomadic people who supplemented their income through brigandage and the trade and transport of frankincense, myrrh and spices. In this early phase of their history, the Nabataeans apparently resisted settling in permanent villages and constructing permanent dwellings. According to Diodorus Siculus, who based his description of the Nabataeans on the writings of Hieronymus of Cardia in the 4th century BC:

> "It is their custom neither to sow corn, plant fruit-bearing trees, use wine, nor construct any house; if anyone is found acting contrary to this, death is his penalty. They follow this custom because they believe that those who possess these things are, in order to retain the use of them, easily compelled by the powerful to do their bidding." (Diod. XIX, 94, 3-4).

One site, described as a "certain rock", possibly the mountain of Um al-Biyara overlooking Petra, was originally used as a stronghold for possessions and family members during annual exchange fairs that were held elsewhere (Diod. XIX, 95, 1-2). According to these early sources Nabataean control of the desert region of the Negev southeast of the southern coast of Palestine was maintained primarily due to the Nabataeans' skill at constructing and hiding cisterns (Diod. XIX, 94, 6-8). A second source presented by Diodorus, that of Agatharchides of Knidos dated to the second century BC, provides descriptions of sites situated on the coast of the Red Sea and in Sinai. This source mentions two peoples, the Maranites and the Garindans, who are described as merchants who transported goods from Sinai to the Gulf of Elat while the Nabataeans are described as Arabs, a numerous people that populated many villages and possessed many flocks (Diod. III, 43,4-6; Dahari 2000:7).

Archaeological research has confirmed much of what is known about the Nabataeans from historical sources. Excavations in the ez-Zantur area of Petra have revealed evidence of tent encampments dating to the late 2nd-late 1st centuries BC underlying stone houses (Schmid 1996: 14-17). Traces of the beginnings of Nabataean settlement activity have been found at Petra from around 100 BC, but permanent structures from that period are quite rare (Schmid 2002: 48). As a rule very few permanent dwellings other than forts have been found in the Negev from before the end of the 1st century BC. It appears that in the Hellenistic period sites in the Negev were probably used on a seasonal basis as campgrounds. This appears to be the case at Oboda, located in the central Negev Highlands, where extensive occupational remains have been found in the form of campsites directly over the bedrock (Negev 1977: 547, 1997: 1, Erickson-Gini 2002:117). At other sites in the Negev Nabataean forts have been found along primary roads and particularly along the Darb es Sultan where forts constructed and occupied in the 3rd-2nd century BC are located at Moa, 'En Rahel and 'En Ziq (Cohen 1981: 36-48, Israel and Nahlieli 1982: 35, Cohen 1984: 25-26). A private dwelling and an aqueduct dating to the 3rd-2nd century BC were unearthed at the site of 'En Rahel on the plain below the fort (Erickson-Gini and Israel 2003: 9*).[1]

[1] Early Nabataean forts have also been found in the Arava Valley at 'En Tamar, located on the southwest edge of the Dead Sea, at Be'er Menuha opposite Wadi Musa and in the Negev Highlands at Horvat Ma'agora

Until around the end of the 2nd century BC Nabataean activity in the Negev in the Hellenistic period appears to have been directly related to the procurement, transport and trade of aromatics from the Arabian peninsula and bitumen from the Dead Sea. Aromatics were purchased from southern Arabians and transferred by way of camel caravans travelling along desert roads through northern Arabia and through the Negev to their destination at the port city of Gaza on the Mediterranean coast.[2] Secondary routes were used to transport goods directly to Egypt, via the Nabataean sites of Nessana and Qasrawit. The latter was located only 25 km southeast of the major Egyptian port at Pelusium in North Sinai (Oren and Netzer 1977: 104). By the late 2nd century BC Nabataean commercial influence was widely felt and ambassadors were sent to Petra from the city of Priene in Asia Minor prior to 129 BC (Bowersock 1983: 22).

The Nabataeans also acted as middlemen trading raw materials in the form of incense resins and spices to third parties. The processing of unguents from raw materials took place at Alexandria and the Greek islands in Mediterranean (Johnson 1987: 36-37). In this period, fusiform unguentaria were used to transport perfumed oils produced in the Mediterranean basin. The fusiform unguentarium or 'spindle bottle' is extremely rare on Nabataean sites (Schmid 2004: 421). On Hellenistic Nabataean sites small alabaster "bee-hive" shaped jars and lids, produced in Yemen, have been discovered that were apparently used to transport unrefined incense resins. Alabaster jars and lids of this type have been found at Petra as well as Nessana and in the central Arava at Moa,'En Rahel and 'En Hazeva[3] (Horsfield 1942: fig. 15: 143, 145; Colt 1962: 60, Pl.XXVI.1-3, 5; Erickson-Gini f.c.).

The lack of permanent Nabataean settlements in southern Jordan and the Negev in this period appears to be related to their lifestyle, much of which was still based on herding and the raising of livestock. Since nearly all the Hellenistic Nabataean sites found in the Negev were located along roads it appears that the Negev was not utilized by the Nabataeans other than as a physical barrier to be crossed by trade caravans during particular seasons. In the Hellenistic period the Nabataeans appear to have been firmly based in the area east of the Arava Valley. This picture receives further confirmation from the apparent cessation of Nabataean trade through the Negev around the end of the 2nd century BC. A hiatus in Nabataean presence has been detected in the Negev sites that may be attributed to the conquest of the important commercial centre of Gaza on the Mediterranean coast by the Hasmonaean king, Alexander Jannaeus (Negev 1997: 1).[4] The conquest of this seaport appears to have disrupted trade, producing an immediate and pronounced effect on Nabataean trade through the Negev. For example, the sites formerly frequented by the Nabataeans were abandoned until the close of the 1st century BC.[5]

The disruption of trade through the Negev occurred in a formative period in Nabataean culture. The first Nabataean coins were minted in the period immediately preceding the conquest by Alexander Jannaeus and these coins are frequently found at sites in the period prior to their abandonment (Kushnir-Stein and Gitler 1992-1993). Similarly, the distinctive Nabataean painted ware tradition appears to have begun in this period. Examples of this ware have been found at Hellenistic Nabataean sites throughout the Negev.

In spite of the hiatus of Nabataean activity in the Negev throughout the greater part of the 1st century BC, the Nabataeans were quite active in extending their influence in other regions. Under Obodas and his successor, Aretas III, the Nabataeans were successful in wars against the Hasmonaean king, Alexander Jannaeus and the Seleucid ruler, Antiochus XII, which enabled them to assert Nabataean rule as far north as Damascus for a brief period in the first half of the 1st century BC (*Jos. Ant.* XIII.15.2, *Wars* 1.4.4). Nabataean involvement in the political affairs of Judea took an important turn after the death of Alexander Jannaeus' widow, Alexandra, in 67 BC. The rivalry between her two sons opened the way for the Idumaean governor, Antipater, himself married to a Nabataean woman from an important family and the father of Herod the Great, to directly influence her older son, Hyrcanus, to seek refuge in Petra in the court of Aretas III (Jos. *Wars* I.8.9, *Ant.* XIV.1.4). This gesture resulted in a military alliance between Hyrcanus and the Nabataeans against his brother, Aristobulus, (Jos. *Ant.* XIV.2). In this period Rome extended its control into the region, effectively turning Nabataea into a client-state after the Roman governor of Syria, M. Aemilius Scaurus, attempted to take Petra in 62 BC (*Jos. Ant.* XIV.5, *Wars* I.8.1). The Nabataean king, Aretas III, bought off Scaurus and prevented the Roman conquest of his kingdom. Coins

(Cohen 1983a: 31.Cohen 1983b: 68-69, Cohen 1985: 76-78, Cohen 1988-1989: 64-65). A fort with round towers discovered at Nessana was probably a Ptolemaic 3rd century BC fort that was eventually occupied by the Nabataeans (Colt 1962: 13). At least two Iron Age sites appear to have been reoccupied by the Nabataeans in this period most notably at Qasr Ruheiba located between Elusa and Nessana and 'En Hazeva in the central Arava Valley (Cohen 1976: 60). Occupational remains in the form of pottery and coins dating to this period have been found at Elusa (Halutza) located along the main Petra – Gaza road (Iliffe 1934: 132-134). The earliest known Nabataean inscription was found at Elusa by Woolley and Lawrence in 1914 (Woolley and Lawrence 1914-1915: Fig. 59).

[2] The Zenon papyri (3rd century BC) refer to the importance of the port of Gaza in the incense trade as well as the involvement of Arabian peoples: the Minaeans and Gerrhaeans and a certain 'Moabite' called Malichus, (a Nabataean name), (PCairo Zen. 59009).

[3] Remains of incense resin were found on a small limestone altar buried in the earliest levels of the fort.

[4] Numismatic evidence shows an increase in the output of the mint at Gaza in the second half of the 2nd century BC indicating a commercial revival in the city before its conquest by Alexander Jannaeus around 96 BC. This commercial revival was probably due to the Seleucid decline and the increased economic activity of the Nabataeans in the late 2nd century BC (Rappaport 1970: 79).

[5] Hasmonaean coins found at Nessana suggest that the site may have been occupied by Hasmonaean forces under Alexander Jannaeus (Bellinger 1962: 70, Urman and Harpak 1994: 49-50). 1st century BC Judaean pottery has also reportedly found at the site (D. Urman, p.c.).

issued by this same king show that he took the title of "Philhellene" in order to emphasize his love for Greek culture (Meshorer 1975: 86-87, Nos. 5-8).

The Monumental Development of Petra and Judaea

In the latter half of the 1st century BC the Greek geographer, Strabo, described the private dwellings of Petra, built with costly stone (Strabo XVI, 4, 26). The development of monumental architecture in Petra in the second half of the 1st century BC, possibly during the reigns of Obodas III (30-9 BC) and Aretas IV (9 BC-AD 40), included temples, such as the Qasr al-Bint al-Firaoun and the Temple of the Winged Lions, as well as the huge complex of the Southern or Great Temple[6] and the *paradeisos* (pool complex), public baths, a monumental gate, a theatre that seated as many as 6,000 people and luxurious private dwellings (Schmid 2002: 49-51, Joukowsky 2002: 451, Bedal 2001: 30-31). Researchers have already noted the influence of Herodian architectural developments on those of Petra (Hammond 1996: 556, Roller 1998: 254). The recent discovery and investigation of the monumental pool complex or *paradeisos* in Petra shows that it is identical to the Herodian *paradeisos* constructed on a grander scale at Herodium in Judaea, leading the excavator to suggest that the Petra complex was modelled on the Herodian predecessor (Bedal 2001:37). In addition, a number of monumental tomb facades and particularly the Khazneh appear to have been carved in the second half of the 1st century BC (Schmid 2002:53).

The monumental construction projects initiated in Nabataea were paralleled with, and apparently influenced by, the monumental building activity of Herod the Great in Judaea. Throughout the remainder of the 1st century BC the fortunes of the Nabataeans were invariably intertwined with those of Judaea and particularly with Herod. Accounts of the period by the Jewish historian, Flavius Josephus, describe an ambivalent relationship between Herod and the Nabataeans that included periodic conflict with occasional references to mutual aid and financial dealings (Jos. *Wars* I.14.1). After Herod's claim to the throne of Judaea was fully established by around 30 BC he initiated an extensive building program inside his kingdom and beyond its borders. During his rule, Herod founded seven cities and undertook major construction projects including the construction and reconstruction of a number of palaces and strongholds in the Judaean Desert such as Jericho, Masada and Herodion, as well as huge projects in Jerusalem including his own fortified palace, the Antonia fortress and the Temple of Jerusalem, the latter, according to Josephus' report, built at Herod's own personal expense (Jos. *Ant.* XV.11.1).

The source of Herod's great wealth, which enabled him to carry out such extensive projects, has never been fully explained. Herod's wealth apparently increased tremendously in the years following the deaths of Cleopatra and Mark Anthony in spite of a number of natural disasters, such as the devastating earthquake of 31 BC and later in the mid-20s when a serious famine struck Judaea, forcing Herod to sell off his own treasures in order to buy imported grain (Jos. *Ant.* XV.9.2). In spite of these setbacks, and his impressive investments in construction projects, by 20 BC Herod reduced taxes by one-third and at the time of his death in 4 BC he was still able to leave huge amounts of money in his will, the sum of which indicates that his estate may have been at least one-third of that of the Roman emperor, Augustus (Jos. *Ant.* XV.10.4, XVII.8.1, Roller 1998: 122).

Herod's wealth was probably directly related to his astute financial investments and profits derived from the production and marketing of perfumed oils. Demand for incense and exotic perfumed oils increased dramatically towards the end of the 1st century BC. These products were used for religious functions, even in the military, as well as for funerals and even for everyday use by those who could afford them (Pliny *NH* 35.5). Herod had a monopoly on the cultivation, processing and marketing of balsam or opobalsam, one of the most expensive products in the ancient world, that was grown exclusively in the area around the Dead Sea (Jos. *Ant.*XIV.4.1, XV.4.2). For a brief period early in his reign, Herod was ordered by Mark Anthony to pay the Egyptian queen, Cleopatra, 200 talents a year to "rent" his own plantations (Jos. *Ant.* XV.4.1, 4; *Wars* I.18.5). The plantations and production centres around the Dead Sea were confiscated by the Romans in the early 1st century AD after Herod's death, and 'En Gedi became the centre of production as part of the emperor's *patrimonium* after 6 AD (Cotton 2001: 142-146). Production facilities operating further south at 'En Boqeq in the first half of the first century AD may have constituted part of the imperial *fiscus* in this period (Cotton and Eck 1997). The fortunes of the Herodian dynasty appear to have been reduced after they were deprived of this source of income. Within the fifty years following the death of Herod the family's wealth was exhausted (Roller 1998: 123-124).

The interest of the Romans in controlling and expanding the cultivation and processing of balsam in the Dead Sea area is readily apparent from historical sources. In the first century AD, Pliny reports that originally balsam was cultivated only in royal plantations at Jericho and 'En Gedi (*NH* 12.54, 111-113) but he claims that in his day it was quite plentiful, an indication that it may have been cultivated elsewhere in the region, possibly as far south as 'En Boqeq and the Nabataean controlled area of Zoara. Pliny also mentions that the plant was exhibited in Rome in the victory procession of Vespasian and Titus following the First Revolt and that during the revolt the Jews attempted to destroy the plants in an effort to damage Roman interests in the region (*NH* 12.54.111-

[6] The identification of this structure as a temple is still in question. The addition of an open-air *theatron* in Phase II of the structure, as well as the lack of an altar, dedicatory inscriptions or other elements associated with temples, raises the possibility that it may have served as a bouleuterion or civic structure for popular assemblies (Schluntz 1999: 82-91, Joukowsky 2001: 453-454).

112, 114).[7] In the period following the revolt the sale of Judaean balsam trimmings alone was worth around 800,000 sesterces (*NH* 12.54, 118), (Hepper and Taylor 2004: 43).

It has recently been pointed out that Pliny's report of balsam plants being cultivated like vines covering whole hillsides accurately describes the remains of ancient agricultural terraces found on the hillsides around 'En Gedi (Magness 2005). Prior to the development of modern 'En Boqeq the area below the springs was extensively terraced, in a manner similar to 'En Gedi. These terraces probably date to the late 1st century BC through mid-1st century AD, during which the only permanent settlement existed at 'En Boqeq. Archaeological research has uncovered possible processing facilities located in Jericho (Bar-Nathan 2002: 51), 'En Boqeq (Fischer *et al.* 2000) and 'En Feshkha (Hirschfeld 2004: 44, 64-65).[8] Significantly, the increased cultivation and production of balsam at Judaean sites next to the Dead Sea appears to be connected with the production of Judaean perfume juglets during the reign of Herod and throughout the first century AD (Fig. 1).[9] This type of juglet, Bar-Nathan's J-JT1A, has been described as "the most common juglet in Judea in the Jerusalem region and in the Herodian palaces at Herodium, Masada and Jericho especially in contexts dating from the First Revolt (66-73 C.E.)", (Bar-Nathan 2002: 163-164).[10]

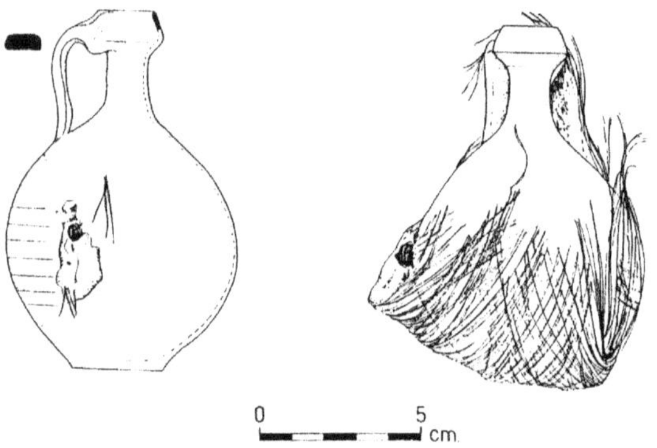

Fig. 1. Juglet from the Judaean Desert (Patrich and Arubas 1989)

[7] In light of this report, the massive assault of the Roman army on nearby Masada may be taken into context. The Romans could have ill afforded the presence of armed rebels located in the heart of this important economic region.

[8] Perforated perfume juglets were found in first century AD contexts in the industrial complex at Jericho (Bar-Nathan 2002: 51). The nature of the site of 'En Feshkha and the function of its industrial complex are still being debated (see Magness 2002b: 210-225; Magness 2005). The *officina* located at 'En Boqeq appears to have functioned in the 1st century AD and as late as the Bar Kochba Revolt (132-135), (Fischer *et al* 2000:3). Although the excavators have suggested that the facility was operated by Jews living at the site, the evidence of Nabataean presence there has prompted other researchers to suggest part of the population at the site were also Nabataean (Magness 2002a: 346-347).

[9] A juglet (Fig. 1) found in a cave near Qumran may have contained balsam oil produced in the region (Patrich and Arubas 1989).

[10] The abundance of this type of juglet in contexts dated to the First Revolt may be coincidental and due to the fact that several instances of destruction layers dated to the revolt have been uncovered in the region of Judaea.

Roman Economic Initiatives and the Nabataean Response

Roman interest in dominating the eastern trade in exotic goods appears to have been the motivation behind orders by the Emperor Augustus to launch an expedition against South Arabia around 26-25 BC. According to Strabo, this expedition was carried out by Strabo's personal friend, the Roman governor of Egypt Aelius Gallus, who led an army of 10,000 men to South Arabia, with contingents provided by the Nabataean king, Obodas III, and Herod of Judaea (Strabo XVI.4.22). The campaign was a failure but it appears to have prompted the beginning of diplomatic relations between Rome and the trading nations of South Arabia and India (Sidebotham 1989: 494). Augustus was the first Roman ruler to receive ambassadors from India and during his reign numerous embassies were dispatched to Rome by the heads of the trading nations of Arabia and India (Casson 1989: 36-38).

By the later part of the 1st century BC, the procurement of exotic goods was facilitated by use of the monsoon winds, the discovery of which was capitalized on by the Romans. Under Augustus, ports, military installations and roads were developed in the Eastern Desert of Egypt (Fig. 2). Roman army units protected the overland caravan routes in the Eastern Desert and the *hydreumata,* or fortified stations with wells or cisterns, that provided accommodation for travellers and pack animals (Sidebotham 1989: 492). Transport by sea and the Nile was faster and less expensive than overland transport through South Arabia. According to Pliny, the overland route consisted of 65 stages that would have taken at least 70 days of travel as compared to only 54 days from Southern Arabia to Alexandria (Pliny *NH* 6.26, 102-104).

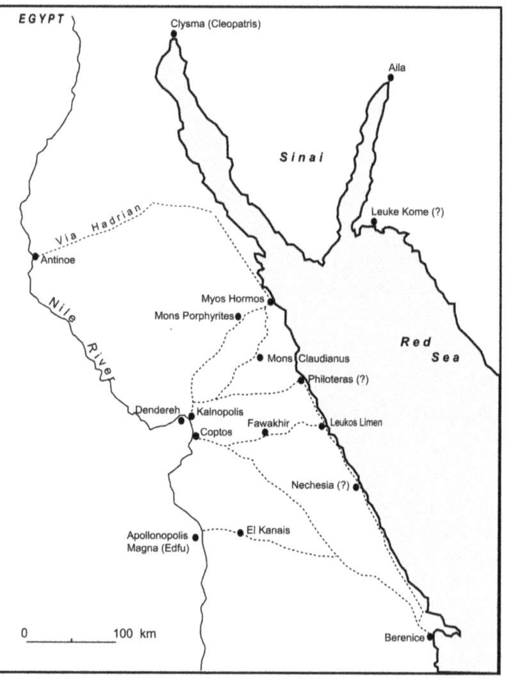

Fig. 2. Roman Trade Routes in the Egyptian Eastern Desert, 30 BCE – 217 CE (Sidebotham 1986: Map 1)

During the reign of Augustus over a hundred ships per year sailed from Egypt's Red Sea ports of Myos Hormos and Berenike to India, six times as many as under the last Ptolemies (Casson 1991: 203).

The Romans paid for exotic spices and aromatics not only in hard currency but also by trading an impressive variety of goods and primarily different types of textiles produced in Egypt and other eastern provinces. The volume of Roman trade with East Africa, Southern Arabia and India was impressive even by modern standards and it appears to have been considerably more successful than that of European trading companies in the early modern period (Raschke 1979: 71).

The Roman trade with India was for large-scale operators, ship-owners and traders, and it was open only to those with large amounts of money and goods needed to obtain expensive eastern imports (Casson 1984: 192-193). Ships bound for India probably departed from Egypt in July in order to take advantage of the southwest monsoon. The Roman ships, built to withstand the force of the monsoon winds, were massive by ancient standards and they were capable of shipping between 350-500 tons on long distance hauls (Raschke 1979: 69). Due to the increased risks of crossing the Indian Ocean these trading ventures required a considerable investment of capital. However, it took only one year to yield returns (Casson 1984: 192). Wealthy merchants, such as Nabataeans, Palmyrenes and Alexandrian Jews, were active in Egyptian commerce and there are indications that even Roman emperors were involved in the financing of trade through their agents (Sidebotham 1986:83, 1989: 496). Inscriptions on ostraca found in Egypt indicate that for sixty years, between the late 1st century BC and the 1st century AD, the Nicanor family operated a transport business between the Red Sea ports of Berenice and Myos Hormos through the Eastern Desert and the emporium at Coptos (Sidebotham 1989: 496-497).

Following the Roman conquest of Egypt its political reorganization apparently had fiscal goals at heart and Egypt came under the personal control of the Emperor Augustus. In this period Augustus reformed the Equestrian Order and changed it into a financial aristocracy in charge of the direct collections of custom duties. The Romans collected impressive amounts of revenue through custom duties (*portoria*) imposed on imports and exports of the eastern trade, and goods imported into Egypt were taxed at 25 percent. Processed goods leaving Alexandria were also taxed but at a lower rate (Miller 1969: 224-225).

Faced with growing competition with Roman Egypt, as well as the rapid development of trade through the Persian Gulf and Palmyra, the Nabataeans reacted by extending their own commercial ventures through the development of new trade routes. The primary innovation in this period appears to have been the genesis of the processing of perfumed oils by the Nabataeans at Petra as proposed by David Johnson (Johnson 1987). Like Herod, the Nabataeans appear to have entered the ranks of a limited number of producers of perfumed oils and thereby vastly increased their profit margin. Evidence for this new economic activity is indirect. However, in the late 1st century BC the Nabataeans began producing their own unguent bottles at Petra and Nabataean unguentaria continued to be produced there until the collapse of international trade in the 3rd century AD (Fig. 3). These bottles have been found, often extensively, in all Nabataean-controlled territories, as well as outside of Nabataea and particularly in Judaea where these vessels have also frequently been found in tombs of the first half of the 1st century AD (Kahane 1952). Nabataean unguentaria have also been found in Europe and Asia Minor at Racatu and Poianna in Dacia (Glodarius 1976: 171, Pl.171; 175, Pl.27), in Rome (Vermaseren and Von Essen 1965: 87), at Stobi in Yugoslavia (Wesolowsky 1973: 120), at Argos (Bruneau 1983: 517) and at Dura

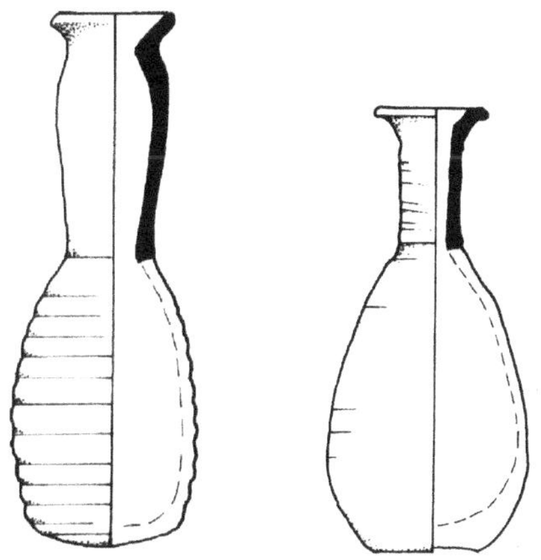

Fig. 3a. Nabataean piriform unguentaria, 1st century AD – 'En Hazeva

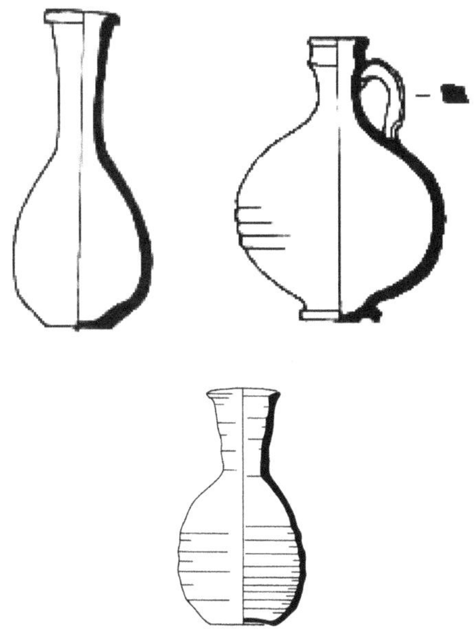

Fig. 3b. Nabataean piriform unguentaria and perfume juglet, 1st century AD – 'En Rahel fort

Europos (Dyson 1968: 11, no. 23).[11] Pottery production in Petra apparently began in the 2nd century BC if not earlier. At first, the Nabataeans produced pottery indistinguishable from other Hellenistic wares but by the second half of the 1st century BC, they began to produce their own distinctive "Nabataean" fine ware pottery found throughout the realm of Nabataea (Schmid 2001: 409). This latter development appears to have coincided with the production of Nabataean ceramic unguentaria.

Johnson has pointed out that dated Nabataean tomb inscriptions at Medain Salih, the centre of Nabataean trade activity in northern Arabia, indicate that this site was visited almost exclusively in the early spring when Nabataean caravans made their way southwards in time for the spring harvest (Johnson 1987: 31-32). According to Pliny the demand for frankincense increased to such an extent that a second crop was introduced (Pliny *NH* 12.32.58). This second crop, harvested in the spring and considered to be of an inferior quality, appears to have been grown for customers involved in the overland caravan trade, and more specifically for the Nabataean processing industry at Petra (Johnson 1987: 30-31).

The development of a protected road system was a vital step in the marketing of Nabataean products. In North Arabia the Nabataeans appear to have operated at least three major routes. The first of these was the inland route connecting South Arabia with Petra by way of Medain Salih. A second route ran along the eastern side of the Red Sea coast, and the third connected the Wadi Sirhan with the Persian Gulf (Johnson 1987: 90). The Red Sea port of Leuke Kome was the only Nabataean seaport through which South Arabian and eastern imports could be shipped, but the development of this port probably took place in order to take advantage of the collection of tax revenues there (Eadie 1989: 117).

Archaeological surveys in the area west of Petra along the Naqb Sleisel route have revealed evidence that the roads leading to the site were part of a planned and organized government project on the part of the Nabataeans. The presence of towers constructed next to this road suggests that movement into Petra was closely watched and controlled (Kloner 1998: 131). In the Negev, the Nabataeans developed the road between Petra and Gaza, creating a new shorter leg between Moa and Oboda by way of the Ramon Crater (Fig. 4). This route was rarely used before the 1st century BC probably due to the difficult terrain and the lack of water sources between Moa and the Ramon Crater. Sometime in the late 1st century BC the Nabataeans created a new pass, the Mahmal Pass, located along the north face of the crater, as well as a series of caravan stops and cisterns along the road. This new road could facilitate intensive year-round transport of goods between Petra and the Mediterranean.

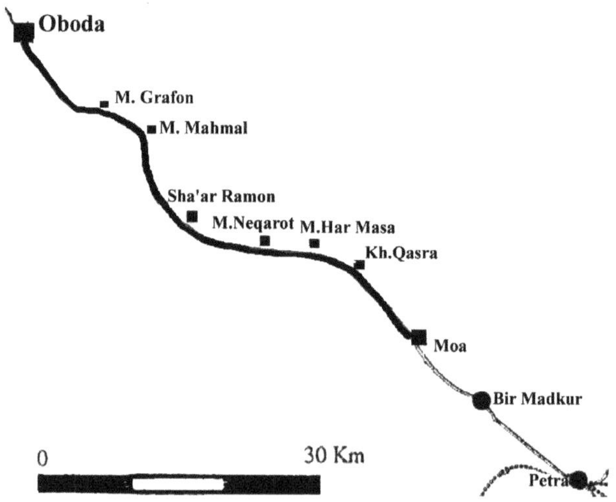

Fig. 4. The Petra – Gaza Road by way of the Ramon Crater 1st - 3rd centuries CE

Judging from the archaeological and historical evidence, it appears that the Nabataeans had adopted dwelling in permanent lodgings and settlements in Petra and possibly other Nabataean sites in the second half of the 1st century BC. This settlement activity, along with the appearance of monumental architecture and pottery production, appears to have accelerated towards the end of that century during the period when the Nabataeans were expanding their political and commercial influence.

Determining the Causes of Nabataean Sedentarization

In earlier research on the origin of Nabataean sedentarization the phenomenon was attributed to the supposed decline of Nabataean trade and the adoption of agriculture as an alternative subsistence strategy in the late 1st century AD (Bowersock 1983: 73-74, Negev 1996: 86). This view was based on Avraham Negev's research in the Negev. Negev proposed that, in spite of the evidence of monumental construction at Petra and the development of roads, forts and caravanserais in the Negev, the Nabataeans continued to live in tents until after the mid-1st century AD. According to him the Nabataeans did not become sedentarized until they switched from trade to agriculture as their economic base during the reign of their last king, Rabbel II (70-106), (Negev 1977: 682, 1988: 13).

More recently, Burton MacDonald has proposed that improved climatic conditions in the early 1st millennium

[11] In a recent article S. Schmid has described Johnson's proposition that these vessels originated in Nabataea as a "somewhat naïve observation" (Schmid 2004: 421). However, there is little doubt that unguentaria were produced in Petra. Kharieh 'Amr has recently described the discovery of "hundreds of unguentaria" in a dump containing Nabataean wares at the site of Gaia in Wadi Musa outside of Petra ('Amr 2004: 238) and Nabataean unguentaria are a significant part of nearly every pottery assemblage found at Nabataean sites. To date, no Nabataean unguentaria have been included in petrographic studies of Nabataean pottery, the emphasis of which has been primarily on the origin of Nabataean painted wares, conducted by 'Amr (1987), Gunneweg *et al* (1988), Bedal (1998). Scientific analysis is required in order to trace the distribution of the unguentaria produced in Petra. Such a study could be a major contribution to our understanding of the Nabataean economy and trade relations.

steady state ⟶ chaos ⟶ bifurcation ⟶ phase transition ⟶ steady state

(Portugali 2000: 318)

AD may have been responsible for the appearance of settlements in the Negev and southern Jordan in the Nabataean, Roman and Byzantine periods (MacDonald 2001). MacDonald's proposal is based on possible evidence of a phase of increased humidity after 190 BC, peaking around 90 AD, as is indicated by studies of the levels of the Dead Sea in historical times (Frumkin *et al.* 1991, Bruins 1994: 307-308).[12] However, Negev's early date for the introduction of intensive agriculture and installations that are found throughout the central Negev has since been challenged by a number of researchers (Haiman 1989, Nevo 1991, Haiman and Fabian in press). Moreover, the development of Nabataean society and settlement appears to have been coincidental with regard to improved climatic conditions. Bruins has pointed out that intensive settlement activity in the Negev in historical times appears to have taken place independent of phases of increased moisture, for example in the very dry period MBI / EBIV (2300-1550 BC) and in the Byzantine and Early Islamic periods (4th-9th century AD). Notably, the humid phase that occurred after 850 AD was not accompanied by the resettlement of the central Negev (Bruins 1994: 310-311).

Archaeological research carried out in the Negev in recent years has revealed evidence showing that, contrary to Negev's claims, trade along the Petra – Gaza road through the Negev continued without interruption until the early 3rd century AD (Cohen 1982: 246, Erickson-Gini and Israel 2003:13*). Permanent dwellings dated to the 1st century AD have since been excavated at both Oboda and Mampsis (Erickson-Gini 1999:11, 2001a, 2001b). This evidence correlates with evidence of sedentarization in Petra in the second half of the 1st century BC (Schmid 2002: 50-51). It now appears that the Nabataeans continued to profit from international trade and the marketing of unguents produced at Petra long after the Roman annexation of Nabataea in 106 AD and until the early to middle 3rd century AD (Johnson 1987: 74, Fiema 2002: 65). The adoption of agriculture on an intensive scale by the inhabitants of the Negev and southern Jordan took place in the 4th century AD, after the disruption of international trade in the wake of the chaotic conditions of the 3rd century (Erickson-Gini, 2005). In light of these findings, it is apparent that Nabataean sedentarization took place in the 1st century BC and that it was not related to the adoption of agriculture as proposed by Negev, or improved climatic conditions as proposed by MacDonald.

Many aspects of Nabataean sedentarization appear to have been the result of acculturation. According to Schmid, as a nomadic people with no preceding material culture, the Nabataeans adopted a foreign culture that included features of "Hellenization" found throughout the Near East in the last centuries of the 1st millennium BC, namely the minting of coins, the production of Hellenistic styles of pottery, sculpture and monumental tomb architecture (Schmid 2001: 408-414, 417). Schmid gauges that the Nabataeans adopted a fully Hellenistic culture around 100 BC and became increasingly open to Roman cultural influences, a development that led to an indigenous Nabataean style of their material culture (Schmid 2001: 415-416).

Through their trade contacts the Nabataeans were exposed to Hellenistic influences as early as the 3rd and 2nd centuries BC. In the next century they took an active part in the dramatic exposure of the East to the Roman Empire. Can cultural exposure, or acculturation, be solely responsible for the transformation of Nabataean society in the 1st century BC?

Nabataean Sedentarization as a Product of Emergent Behaviour

One possible explanation for the phenomenon of Nabataean sedentarization is that it was the result of a process of emergent behaviour, or self-organization, of a complex adaptive system.[13] This process was first documented in the natural sciences and has since been increasingly applied to social and cultural changes (De Landa 1997, Portugali 2000, Johnson 2001). In their studies on the transformation of Nabataean culture both Schmid and Johnson have detected this process at work. According to Johnson, rapid sedentarization of the nomadic Nabataeans took place when they shifted from part-time trading activities to a full-time specialization in the incense trade and particularly the production of unguents in Petra itself (Johnson 1987: 143). Both Johnson and Schmid propose that Nabataean sedentarization was a

[12] In their critique of the use of environmentally deterministic models in archaeology Rosen and Rosen point out that in cases in which climatic changes have been assumed to correlate with social changes the models used by researchers are often fundamentally simplistic and they provide little information about the events that influence human behavior (Rosen and Rosen 2001: 538-539). Moreover, Magness has highlighted the problematic nature of the use of environment and climate in regards to settlement in the Dead Sea region in the early first millennium C.E. and she points out that the remains of permanent, multiperiod settlements on its western shore have only been found at Jericho and 'En Gedi, oases with perennial sources of fresh water (Magness 2005).

[13] Originally the study of self-organization began in the field of thermodynamics with a theory that suggests that order can take place spontaneously when systems are pushed far from a state of equilibrium. According to the principle of self-organization, new structures arise at bifurcation points, or thresholds of instability.

response to new economic conditions in which the Nabataeans were forced to create a more permanent infrastructure in order to remain competitive in trade (Johnson 1987: 21-33, Schmid 2001: 415).

In terms of the self-organization process, this rapid transformation from pastoralism to sedentarization may be termed a "phase transition". The Nabataeans' former nomadic lifestyle appears to have been a stable one in which they subsisted on pastoralism for several centuries and augmented their income through seasonal trade, acting as middlemen in providing transport for incense resins obtained from South Arabia, that were sold in the Mediterranean basin. The stability of this mode of subsistence was disrupted in the 1st century BC with the expansion of the Roman Empire and the increased demand for aromatics, and consequent increased competition in the aromatics trade.

Social transformations of this type, or "phase transition" in the language of self-organization, occur when old regimes disintegrate and enter into a chaotic period or period of fluctuation. This stage is necessary in order to enable the emergence of new forces and orders that bring the system into a new steady state or mode of production (Portugali 2000: 318).[14]

Based on Portugali's model above, it is possible to create a new model for the transformation of Nabataean culture and society in which it is apparent that the Nabataeans responded to economic realities by adopting a new form of subsistence: the production and marketing of unguents. The production and marketing of unguents required a more sophisticated social and political organization and largely sedentarized lifestyle. These activities provide a ready explanation for the expansion of Nabataea in the later 1st century BC and the 1st century AD, when new roads were built and permanent service colonies were established outside of the Petra area, as can be seen in the central Negev. Another by-product of this industry was increased pottery production, since locally produced unguentaria were required to store and ship processed perfumed oils.

As I have shown above, the production and marketing of unguents by the Nabataeans in the late 1st century BC was not unprecedented. Their nearest neighbour, Herod king of Judaea, controlled the cultivation, processing and marketing of expensive balsam oil grown exclusively around the Dead Sea on the border of Nabataea. The Roman emperor Augustus consolidated Egypt and built new roads and stations in order to take over the transport of exotic goods and particularly incense resins obtained in Southern Arabia and India, as well as the processing centres at Alexandria. In light of the accelerated volume of production and trade in the Roman East in this period it is not surprising that the Nabataeans took part. As Rosen has pointed out, cultural florescence in the Negev region in this and other historical periods appears to be related to increased economic and social ties with the Mediterranean core area (Rosen 2000: 47).

The prosperous trade in unguents and other exotic goods poured vast amounts of revenue into the coffers of Rome, Judaea and Nabataea. For the Nabataeans, the increased flow of money, which may be compared to an increased flow of energy into their society (Odum and Odum 1981: 41), acted as a catalyst for rapid social and cultural change. Although at present there is a paucity of material or historical evidence documenting Nabataean business in the form of receipts and contracts, their investment of revenue in the form of elaborate tombs, temples, private villas, a theatre and even a *paradeiso* or garden pool complex at Petra is readily apparent, making that isolated desert city one of the wonders of the ancient world.

[14] Studies carried out on the process of organizational collapse of inter-regional systems of Indian tribes in the American Southwest have revealed that economic and social disruption and instability motivates individuals and communities to experiment with new organizational principles (Stone 1999: 117). In later Nabataean history, the collapse of international trade through the Negev in the mid-3rd century AD led to a new economic regime and accelerated urbanization in the region in the 4th century (Erickson-Gini, 2005).

Bibliography

Classical sources

Diod. Diodorus of Sicily (1st century B.C.E.). 1960-1967, *Bibliotheca historica*, ed. C.H. Oldfather, London: Heinemann.

Diodorus of Sicily, 12 volumes. 1954, transl. R.M. Geer. Cambridge, Mass: Harvard University.

Jos. Ant. Josephus Flavius. 1976, *Complete Works: The Antiquities of the Jews*, transl. W. Whiston, New York: Seabury.

Jos. Wars. Josephus Flavius. 1976, *Complete Works: The Wars of the Jews*, transl. W. Whiston, New York: Seabury.

PCairo Zen. 1925-1940, *Zenon Papyri*, I-V (Catalogue general des antiquities egyptiennes du Musee du Caire), C.C. Edgar *et al.*, nos. 59001-59853, Cairo.

Pl. C. Plinius Secondus, *Naturalis Historia*. 1947-1963, ed. H. Rackham, London: Heinemann.

Strabo *The Geography of Strabo*. 1947-1963, ed. H.L. Jones, London: Heinemann.

Literature

'Amr, K.

1987 *The Pottery from Petra: A Neutron Activation Analysis Study*. Oxford: British Archaeological Reports International Series No. 324.

2004 "Beyond the Roman Annexation: The Continuity of the Nabataean Pottery Tradition". *Studies in the History and Archaeology of Jordan* VIII: 237-245.

Bar-Nathan, R.

2002 *Hasmonean and Herodian Palaces at Jericho: The Pottery. Vol. III.* Jerusalem:Israel Exploration Society.

Bedal, L.
1998 "Neutron Activation Analysis of Pottery" in *Petra Great Temple Volume I: Brown University Excavations 1993-1997.* ed. M.S. Joukowsky. Providence: Brown University. pp. 345-368.
2001 "A Pool Complex in Petra's City Center", *Bulletin of the American School of Oriental Research* 324: 23-41.
Bellinger, A.R.
1962 "Coins" in *Excavations at Nessana*, vol I. H. D. Colt D. London: British School of Archaeology. pp. 70-75.
Bowersock, G.W.
1983 *Roman Arabia.* Cambridge, Mass: Harvard University.
Bruins, H. (1994) "Comparative Chronology of Climate and Human History in the Southern Levant from the Late Chalcolithic to the Arab Period" in Late Quaternary and Paleoclimates of the Eastern Mediterranean, eds O. Bar-Yosef, and R.S. Kra. *Radiocarbon.* Tucson: University of Arizona. pp. 301-14.
Bruneau, P.
1983 "Tombes De Argos". *Bulletin Correspondance Hellenique* 94: 517.
Casson, L.
1984 *Ancient Trade and Society.* Detroit: Wayne State University.
1989 *The Periplus Maris Erythraei.* Princeton: Princeton University.
1991 *The Ancient Mariners. Seafarers and Sea Fighters of the Mediterranean in Ancient Times.* Princeton: Princeton University.
Cohen, R.
1976 "Qasr Ruheibe". *Hadashot Archeologiot* 59-60: 59-60 (Hebrew).
1981 "Moa". *Hadashot Archeologiot* 77: 36-38 (Hebrew).
1982 "New Light on the Date of the Petra – Gaza Road". *Biblical Archaeologist* 45:240-247.
1983a "En Tamar". *Excavations and Surveys in Israel* 2:31.
1983b "Mesad Be'er Menuha". *Excavations and Surveys in Israel* 2:68-69.
1984 "Ein Ziq". *Excavations and Surveys in Israel* 3:25-26.
1985 *Archaeological Survey of Israel, Map of Sede Boqer – West (167)* 12-03. Jerusalem: Archaeological Survey of Israel. Israel Antiquities Authority.
1988-1989 "Horvat Ma'agura". *Excavations and Surveys in Israel* 7-8:114.
Colt, H. D.
1962 *Excavations at Nessana.* Vol. I. London: British School of Archaeology.
Cotton, H. M.
2001 "Ein Gedi Between the Two Revolts". *Scripta Classica Israelica* 20: 139-154.
Cotton, H.M., and Eck, W.
1997 "Ein Staatsmonopol und seine Folgen: Plinius Naturalis Historia 12, 123 und der Preis fur Balsam". *Rheinisches Museum fur Philologie* 140: 153-161.
Dahari, U.
2000 *Monastic Settlements in South Sinai in the Byzantine Period.* Jerusalem: Israel Antiquities Authority.
De Landa, M.
1997 *A Thousand Years of Nonlinear History.* New York: Swerve.
Dyson, S.
1968 *The Excavations of Dura Europus, Final Report IV.* New Haven: Yale University.
Eadie, J. W.
1989 "Strategies of Economic Development in the Roman East: The Red Sea Trade Revisited" in *The Eastern Frontier of the Roman Empire. Proceedings of a Colloquium Held at Ankara September 1988*, eds D.H. French, and C.S. Lightfoot. Oxford: British Archaeological Reports International Series 553 (i). pp. 113-20.
Erickson-Gini, T.
1999 Mampsis: A Nabataean Roman Settlement in the Central Negev Highlands in Light of the Ceramic and Architectural Evidence Found in Archaeological Excavations During 1993 1994. Unpublished M.A. dissertation, Tel Aviv University.
2001a "Oboda; Abde; Avdat. New Excavations". in *Archaeological Encyclopedia of the Holy Land*, eds A. Negev, and S. Gibson. New York: Continuum. pp. 374-375.
2001b "Mampsis. New Excavations" in *Archaeological Encyclopedia of the Holy Land*, eds A. Negev, and S. Gibson. New York: Continuum. p. 312.
2002 "Nabataean or Roman? Reconsidering the Date of the Camp at Avdat in Light of Recent Excavations" in *Limes XVIII – Proceedings of the XVIIIth International Congress of Roman Frontier Studies held in Amman, Jordan (September 2000)*, eds P.W.M. Freeman; J. Bennett; Z.T. Fiema; and B. Hoffmann. Oxford: British Archaeological Reports International Series, vol. I. pp. 113-30.
2005 Crisis and Renewal – Settlement in the Central Negev in the Third and Fourth Centuries C.E, with an Emphasis on the Finds from Recent Excavations at Mampsis, Oboda and Mezad 'En Hazeva. Unpublished PhD dissertation. Jerusalem: Hebrew University.
Erickson-Gini, T., and Israel, Y.
2003 "Recent Advances in the Research of the Nabataean and Roman Negev". in *The Nabateans in the Negev*, ed. R. Rosenthal-Heginbottom. Haifa: Reuben and Edith Hecht Museum, University of Haifa. pp. 9*-14* (English), 11-14 (Hebrew).
Fiema, Z.T.
2002 "From the Annexation to Aaron: Petra in Roman and Byzantine Times" in *Petra: A City Forgotten and Rediscovered*, eds J. Frosen, and Z.T. Fiema. Helsinki: Amos Anderson Art Museum. pp 60-73.

Fischer, M.; Gichon, M.; and Tal, O.
2000 *En Boqeq: Excavations in an Oasis on the Dead Sea. Vol. 2. The Officina, An Early Roman Building on the Dead Sea Shore.* Mainz: Philipp von Zabern.
Frumkin, A.; Magaritz, M.; Carmi, I.; and Zak, I.
1991 "The Holocene Climatic Record of the Salt Caves of Mount Sedom, Israel". *The Holocene* 1: 190-200.
Glodarius, I.
1976 *Dacian Trade with the Hellenistic and Roman World.* Oxford: British Archaeological Reports Supplementary Series, No. 8.
Gunneweg J.; Perlman I.; and Asaro F.
1988 "The Origin, Classification and Chronology of Nabatean Painted Fineware". *Jahrbuch des roemisch-germanischen Zentralmuseums Mainz* 35:1-30.
Haiman, M.
1989 "Preliminary Report of the Western Negev Highlands Emergency Survey". *Israel Exploration Journal* 39: 173-191.
Haiman, M., and Fabian, P.
In press "Town – Countryside Relationships in the Negev Desert in the 5th-8th Centuries C.E." *Proceedings of the Third International Congress on the Archaeology of the Ancient Near East.* Paris: Sorbonne.
Hammond, P.
1996 "Petra" in *The Dictionary of Art*, ed. J. Turner. New York: Oxford University Press. pp. 556-58.
Hepper, F. N., and Taylor, J.E.
2004 "Date Palms and Opobalsam in the Madaba Mosaic Map". *Palestine Exploration Quarterly* 136: 35-44.
Hirschfeld, Y.
2004 "Excavations at 'Ein Feshkha, 2001: Final Report'", *Israel Exploration Journal* 54: 37-74.
Horsfield, G., and Horsfield, A.
1942 "Sela-Petra, The Rock of Edom and Nabatene". *Quarterly of the Department of Antiquities of Palestine* 9: 105-218.
Iliffe, J.H.
1934 "Nabataean Pottery from the Negev". *Quarterly of the Department of Antiquities of Palestine* 3:132-135.
Israel, Y., and Nahlieli, D.
1982 "'En Rahel". *Israel Exploration Journal* 32:163.
Johnson, D.J.
1987 Nabataean Trade: Intensification and Culture Change. PhD Dissertation, Ann Arbor: The University of Utah.
Johnson, S.
2001 *Emergence.* London: Kluwer Academic Publishers.
Joukowsky M.
2001 "Petra: Brown University Excavations of the Great Temple. Questions about Functional Analysis". *Studies in the History and Archaeology of Jordan* VII: 447-455.
2002 "The Brown University 2001 Petra Great Temple Excavations Offer More Surprises". *Annual of the Department of Antiquities of Jordan* XLVI: 315-330.
Kahane, P.
1952 "Pottery Types from Jewish Ossuary Tombs around Jerusalem". *Israel Exploration Journal* 2:176-182.
Kloner, A.
1998 "Stepped Roads in Roman Palestine". *ARAM Periodical* 8:111-137
Kushnir-Stein, A., and Gitler, H.
1992-1993 "Numismatic Evidence from Tel Beer-Sheva and the Beginning of Nabataean Coinage". *Israel Numismatic Journal* 12: 13-20.
MacDonald, B.
2001 "Relation between Paleoclimate and the Settlement of Southern Jordan during the Nabataean, Roman and Byzantine Periods", in *Studies in the History and Archaeology of Jordan* VII: 373-378.
Magness, J.
2002a "Review". *American Journal of Archaeology* 106:346-347.
2002b *The Archaeology of Qumran and the Dead Sea Scrolls.* Grand Rapids, MI: Eerdmans.
2005 "Review". *Review of Biblical Literature* Vol. 7. http://www.bookreviews.org/subscribe.asp.
Meshorer, Y.
1975 *Nabataean Coins.* Qedem Reports 3. Jerusalem: Hebrew University.
Miller, J.I.
1969 *The Spice Trade of the Roman Empire.* Oxford: Oxford University.
Negev, A.
1977 "The Nabateans and the Provincia Arabia" in *Aufstieg und Niedergang er Romischen Welt II.6*, eds M. Temporini, and W. Haase. Berlin and New York: Walter de Gruyter. pp. 520-686.
1996 "Oboda: A Major Nabatean Caravan Halt". *ARAM Periodical* 8:1 & 2: 67-87.
1997 *The Architecture of Oboda, Final Report.* Qedem Reports 36. Jerusalem: Hebrew University.
Nevo, Y.
1991 *Pagans and Herders. A Re-examination of the Negev Runoff Cultivation Systems in the Byzantine and Early Arab Periods.* Beersheba: Midreshet Ben Gurion.
Odum, H.T., and Odum, E.
1981 *Energy Basis for Man and Nature.* New York: McGraw-Hill.
Oren, E., and Netzer, E.
1977 "Settlements of the Roman Period at Qasarweit in Northern Sinai". *Qadmoniot* X. No. 4(40): 94-107.
Patrich, J., and Arubas, B.
1990 "A Juglet with Balsam Oil (?) from a Cave Near Qumran". *Eretz Israel* 21: 321-332.
Portugali, J.
2000 *Self-Organization and the City.* Berlin: Springer.
Rappaport, U.
1970 "Gaza and Ascalon in the Persian and Hellenistic Periods in Relation to their Coins". *Israel Exploration Journal* 20: 75-80.

Raschke, M.G.

1979 "The Role of Oriental Commerce in the Economies of the Cities of the Eastern Mediterranean in the Roman Period". *Archaeological News* VIII, No. 2/3: 68-77.

Roller, D.W.

1998 *The Building Program of Herod the Great.* Berkeley: University of California.

Rosen, S. A.

2000 "The Decline of Desert Agriculture: A View from the Classical Period Negev" in *The Archaeology of Drylands: Living at the Margins*, eds G. Barker, and D. Gilbertson. One World Archaeology 39. New York: Routledge. pp. 45-62.

Rosen, A.M., and Rosen, S.A.

2001 "Determinist or Not Determinist? Climate, Environment, and Archaeological Exploration in the Levant" in *Studies in the Archaeology of Israel and Neighboring Lands in Memory of Douglas L. Esse*, ed. S. M. Wolff. Chicago: Oriental Institute of the University of Chicago. pp. 535-49.

Schmid, S.G.

1996 "Der Feinkeramik" in *Petra: Ez Zantur I: Ergebnisse der Schweizerisch-Leichtensteinischen Ausgrabungen 1988-1992*, eds A. Bignasca; N. Desse-Berset; R. Fellmann-Brogli; R. Glutz; S. Karg; D. Keller; B. Kolb; C. Kramar; M. Peter; S.G. Schmid; C. Schneider; R.A. Stucky; J. Studer; and I. Zanoni. Mainz: Philipp von Zabern. pp 151-218.

2001 "The 'Hellenisation' of the Nabataeans: A New Approach". *Studies in the History and Archaeology of Jordan* VII: 407-378.

2002 "From Aretas to the Annexation: Petra and the Nabataeans" in *Petra: A City Forgotten and Rediscovered*, eds J. Frosen, and Z.T. Fiema. Helsinki: Amos Anderson Art Museum. pp. 44-59.

2004 "The Distribution of Nabataean Pottery and the Organization of Nabataean Long Distance Trade". *Studies in the History and Archaeology of Jordan* VIII: 415-426.

Schluntz, E.L.

1999 From Royal to Public Assembly Space: The Transformation of the "Great Temple" Complex at Petra, Jordan. Unpublished Ph.D. dissertation. Providence: Brown University.

Sidebotham, S.E.

1986 *Roman Economic Policy in the Erythra Thalassa, 30 BC – AD 217.* Mnemosyne Supplement 91. Leiden: Brill.

1989 "Ports of the Red Sea and the Arabia-India Trade" in *The Eastern Frontier of the Roman Empire. Proceedings of a Colloquium Held at Ankara September 1988*, eds D.H. French, and C.S. Lightfoot. Oxford: British Archaeological Reports International Series 553 (i). pp. 485-513.

Stone, T.

1999 "The Chaos of Collapse: Disintegration and Reintegration of Inter-Regional Systems". *Antiquity* 73: 110-118.

Urman, D., and Harpak, T.

1994 "Renewed Excavations at Nessana" in *Nizzana and the Western Border*, eds E. Orion, and Y.Eini. Midreshet Sde Boker. (Hebrew). pp. 49-54.

Vermaseren, J.J., and Von Essen, C.C.

1965 *The Excavations in the Mithream of the Church of Santa Prisca in Rome.* Leiden: Brill.

Wesolowsky, A.B.

1973 "Burial Customs in the West Cemetery" in *Studies in the Antiquities of Stobi.* Vol. 1, ed. D.J. Wiseman. Austin: University of Texas. pp. 98-177.

Woolley, C.L., and Lawrence, T.E.

1914-1915 *The Wilderness of Zin.* London: Palestine Exploration Fund Annual III.

PASTORALISM AND AGRICULTURE IN THE NEGEV IN THE IRON AGE II[1]

Moti Haiman

Introduction

About 350 well-preserved sites from the Iron Age II are known in the Negev Highlands today (Woolley and Lawrence 1915: 61-71; Anati 1955; Glueck 1959; Evenari *et al.* 1958; Aharoni, *et al.* 1960; Aharoni 1967; 1976; Rothenberg 1967: 71-9; Meshel 1974; 1994; Cohen 1981; Haiman 1986, 1991, 1993, 1998, 2003, in press; Lender 1990). These sites include about 60 fortresses, hundreds of dwelling structures and a large number of installations: cisterns, silos, threshing floors and animal pens (Figs 1 & 2).

Most of the studies of these settlements relate to the chronological and historical aspects, as well as to the nature and function of the fortresses – the dominant characteristic of the period, while the environmental aspect has not received sufficient attention.

In this paper I would like to address aspects of environment and subsistence, such as water supply, agriculture and animal husbandry, based on the analysis of installations and finds like sickle blades, which indicates an intensive relationship with the desert environment.

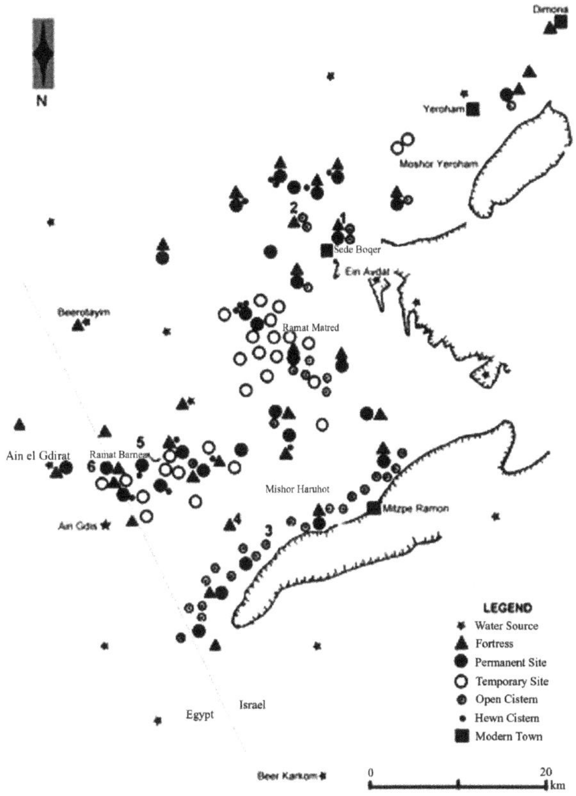

Fig.1. Distribution of Iron Age II sites in the Negev. Sites mentioned in the text or in other figures: 1. Atar Haroah; 2. Hurvat Haluqim; 3. Bor Hemet; 4. Har Gizron; 5. Nahal Mitnan; 6. Wadi el Gdirat

The aspect of water supply

The available water sources in the area, such as Ain el Gdirat, Ain Qadis and others, form a ring around the Negev Highlands. The Iron Age sites constitute a special phenomenon, with a limited distribution in the heart of the region, up to 15 km from a natural water source, and therefore totally depended on cistern water. By comparison, in the Early Bronze II and IV permanent settlements were found next to water sources while temporary settlements were located at various distances from them. Another characteristic of the Iron Age settlements is that their distribution is limited to a territory within range of the fortresses.

The Iron Age sites show a gradual change in size as one moves south. This change is a function of the increasing distance from the margins of the sedentary land to the north and is not related to the presence of natural water sources. The largest settlements in the Negev such as Atar

Fig. 2. The fortress of Har Gizron (No. 4 in fig. 1; photo M. Haiman. Courtesy of the Israel Antiquities Authority).

[1] This paper updates some aspects from a previous article by this author (Haiman 1993).

Haroah and Hurvat Halukim (Cohen 1981 sites 64, 98) are located in the northern distribution area, about 8 km away from a natural water source, while settlements found next to Ain el Gdirat (Fig. 3) and Ein Qadis are much smaller (Haiman, in press).

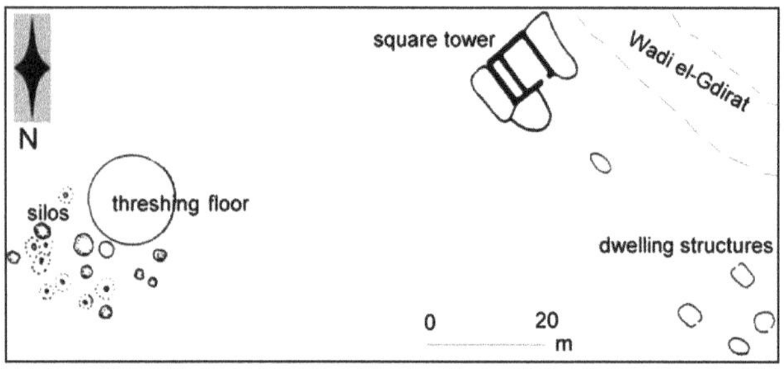

Fig. 3. Site near Wadi el Gdirat (No. 6 in fig 1; photo M. Haiman. Courtesy of the Israel Antiquities Authority)

In the settlement distribution area, about 200 cisterns (of which 100 are open cisterns, Fig. 4) could be associated with Iron Age sites, and these may have supplied the needs of the population.

Fig. 4. Bor Hemet (No. 3 in fig 1) – Open cistern filling (photo M. Haiman. Courtesy of the Israel Antiquities Authority).

Examination of the open cisterns revealed a direct correlation between the number of cisterns associated with a site and the distance of the site to a natural water source. In the area of Sede Boqer, which is located 8 km away from a natural water source, 7 structures were found per cistern (about 75 cu/m per structure). This amount considerably exceeds the annual consumption of a family and its animals (about 20 cu/m). At Ramat Matred and Ramat Barnea, which are located 10 km away from a natural water source, there were 4 structures per cistern, and in Mishor Haruhot, up to 15 km from a natural water source, 3 open cisterns were found per structure.

Theoretically, the open cisterns alone could have supplied enough drinking water for the entire population; however, the actual situation is different.

Due to the unstable rainfall regime of the area, with droughts every other year or every third year, the phenomenon of localized rainfall and the considerable differences in the amount of annual rainfall from one year to the next, the cisterns were not filled, or not filled to full capacity every year. In addition, for the open cisterns the evaporation rate in the area is 2.5 sq m per year per cistern. This puts into question the capability of the open cisterns to hold water beyond the beginning of the summer (Negev 1979). This is critical in those areas that only have open cisterns.

A partial response to the problem of water supply in the area, from the perspective of environmental constraints, can be found in the nature of the settlement. Out of a total of 1,200 structures found in the sites, 20% were of the type well-known in the north (Fig. 5). These were always found near cisterns, and could be considered permanent (amounting to an estimated 250 resident families); 80% of the structures were round and poorly built, mostly without any connection to a cistern: these must be considered seasonal dwellings (Fig. 6).

Fig. 5. Four-room house at Atar HaRoah, excavated by Cohen (No. 2 in fig 1; photo M. Haiman. Courtesy of the Israel Antiquities Authority).

Fig. 6. Round structure in the site near Wadi el Gdirat (See fig 3 and No. 6 in fig 1; near Wadi el Gdirat (See Fig. 3 and No. 6 in Fig 1; photo M. Haiman. Courtesy of the Israel Antiquities Authority).

There was, therefore, a limited number of permanent settlements, while the larger number of temporary sites reflect seasonal activity of farming and herding. The similarity in the material culture of the temporary sites and the permanent sites (with the exception of the architecture of the dwellings), strongly suggests that the seasonal activity was conducted from the permanent sites. However, the problem of water supply in drought years remains (Negev 1979).

The aspect of animal husbandry

About 400 animal pens were found in the sites, showing that the inhabitants were heavily engaged in sheep and goat husbandry. The estimated total floor area of the pens is 45,000 sq m, which could accommodate *ca.* 22,000 heads. Assuming that many of the pens reflect seasonal herding drives from the permanent settlement, the actual number of animals may be a third of this figure, about 7,000 heads or about 20-30 animals per family, which is close to the size of the average herd of present-day Bedouin families in the southern Negev and the Sinai; this herd size is much lower than the minimum of ca. 100 heads that is considered necessary for subsistence (Evenari *et al.* 1971; Marx 1974; Bailey 1977; Abu-Rabia 1994).

The aspect of agriculture

Considerable evidence of agricultural activity was found: an abundance of sickle blades, silos and threshing floors. The silos can be classified into two types:

A) Roofed, bell-shaped silos that widen towards the bottom, the bases ranging between 1-2 m. in diameter, while the upper part is around 1 m. in diameter, and the depth is approximately 2 m (Fig. 7).

Fig. 7. Roofed silo in the site near Wadi el Gdirat (See fig 3 and No. 6 in fig 1; photo M. Haiman. Courtesy of the Israel Antiquities Authority)

B) Open, round or square silos, with a surface opening of 1-2 x 1-4 m. and 1-2 m. deep (Fig. 8).

Both types are dug into the ground and lined by small stones.

Fig. 8. Open silo in the site near Wadi el Gdirat (See fig 3 and No. 6 in fig 1; photo M. Haiman. Courtesy of the Israel Antiquities Authority)

The threshing floors that have been attributed to Iron Age sites consist of a circle of stones, 8-15 m. in diameter, surrounding a leveled compacted dirt floor (Fig. 9). Some of the threshing floors have adjacent silos, indicating that the grain was stored immediately after threshing.

Fig. 9. Threshing floor in an Iron Age II site at Nahal Mitnan (No 5 in fig.1; photo M. Haiman. Courtesy of the Israel Antiquities Authority)

A related issue is the availability of agricultural land, since the only arable land in the area is the wadi bed. Although terraced wadis have been dated to this period (Evenari *et al.* 1971; Bruins 1986; Cohen 1986) surveys show that many Iron Age sites at which the above-mentioned evidence of agricultural activity was found, were not located near terraced wadis. Furthermore, in many areas, such as Shivta, where most of the wadis were terraced, Iron Age settlement is sparse. The terraced wadis in the Negev encompass *ca.* 300,000 dunam, while less than 4,000 dunam would have been required to grow enough grain to sustain the Iron Age population. Moreover, there is a strong contrast between the sophisticated agriculture reflected by the terraced wadi system and the low level of the material culture represented by the Iron Age sites. Taking into account fluctuations in the amount of natural soil in the wadi beds, it appears that, at least until the Iron Age, there was enough natural soil for agricultural cultivation and large-scale terracing of the wadis was not necessary. It seems that the problem did not lie in the amount of land available for cultivation, but rather in growing grain in the desert environment, with its average of 80 mm. annual rain fall.

It is clear that under these conditions: the severe limitations in water supply and the frequency of draught, the inhabitants of the Negev could not support themselves. In addition to the environmental limitations – which prevail in the entire desert – the fact that the settlement distribution is limited to the Negev Highlands – a phenomenon that appeared only in the Iron Age – renders the conclusion of subsistence on local resources impossible. Even seasonal activities like herding and farming, which one would expect to have a much more extensive distribution, in order to increase the limited resources, do not exceed the area of the fortresses. Instead, there is preference for proximity to fortresses over optimal exploitation of the natural resources.

The conclusion, therefore, is that other considerations, not connected with the exploitation of environmental conditions for subsistence, caused this population of farmers and shepherds to migrate into the desert.

Speculations on the Actual Desert Ecosystem

As shown above, there is a contradiction between the location of the Iron Age sites and the location of the natural water sources. The same contradiction exists between the demands of farming and herding in harsh desert conditions, and the fact that these activities were limited only to the area between the fortresses. This settlement location, which seems to prefer proximity to fortresses over vital environmental conditions, suggests the possibility that the settlement system constituted a complex system representing the interest of the state to settle the frontier, in a region where the poor economic activity could support the settlers only to a limited extent.

Studies on the Bedouin economy in these harsh areas show engagement in the traditional occupations of pastoralism and agriculture, but these are of secondary importance. The main source of subsistence is the connection with permanent settlers (Marx 1974; Bailey 1977; Shmueli 1977, 1980).

It is assumed that the Iron Age II settlement reflected the need of the state – the United Monarchy of Israel in the 10th century B.C. – to settle the southern frontier, which was bordered by Egypt and Edom (Bartlett 1980; Malamat 1983). Therefore, the main source of subsistence for the settlers was the support of the state, and the presence of animal pens or sickle blades (Fig. 10) does not necessarily mean full subsistence on agriculture or pastoralism.

The settlement reflected a type of site that appeared in sedentary land to the north at the end of the 11th century B.C., which consisted of a fortress, dwelling structures

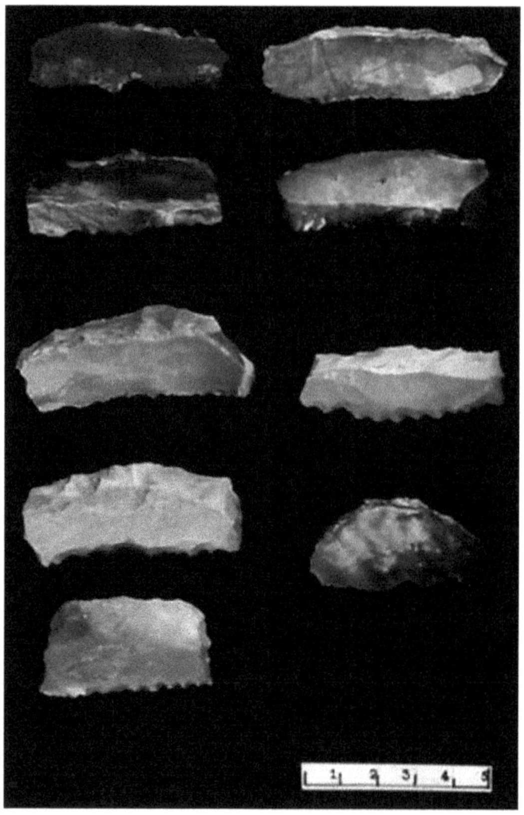

Fig. 10. Collection of Iron Age sickle blades from the Negev (photo N. Sne; Courtesy of the Israel Antiquities Authority)

and agricultural installations (Fritz and Kempinski 1983; Gal 1992; Zertal 1996). This type of settlement continued to exist in the sedentary regions in the north until the end of the Iron Age, primarily in frontier areas (Mazar 1982).

Thus, the impossible settlement in the Negev was a provisional settlement dictated by political rather than environmental considerations and could only have existed as long as the specific geopolitical circumstances existed. When conditions changed following Shishak's campaign (Mazar 1957; Ahituv 1979), the settlements were abandoned and the border moved northward to the Beersheva valley.

Bibliography

Abu-Rabia, A.
1994 *The Negev Bedouin and Livestock Rearing*, Oxford: BERG.
Aharoni,Y.
1967 "Iron Age Fortresses in the Negev". *Israel Exploration Journal* 17:1-17, 97-111.
1976 "The Settlement of the Tribes in the Negev - A New Picture". *Ariel* 41:3-19.
Aharoni, Y.; Evenari, M.; Shanan, L.; and Tadmor, N.H.
1960 "The Ancient Desert Agriculture of the Negev". *Israel Exploration Journal* 10: 23-36.

Ahituv, S.
1979 The Egyptian Topographical Lists Relating to the History of Palestine in the Biblical Period. Ph.D. dissertation, Hebrew University, Jerusalem (Hebrew with English summary).
Anati, E.
1955 *The Central Negev*. Unpublished survey report. Israel Antiquities Authority. (Hebrew).
Bailey, I.
1977 "The Bedouins in Sinai" in *The Desert, Past-Present-Future*, ed. E. Zohar. Tel Aviv: Reshafim (Hebrew). pp. 240-47.
Bartlett, J. R.
1980 *Edom and the Edomites*. Sheffield: Sheffield University.
Bruins, J.H.
1986 *Desert Environment and Agriculture in the Central Negev and Kadesh Barnea During Historical Periods*. Nijkerk, the Nederlands: MIDBAR foundation.
Cohen, R.
1981 *Map of Sede Boqer East (168)*. Jerusalem: The Archaeological Survey of Israel. Israel Exploration Society.
1986 The Settlement of the Central Negev in Light of Archaeology and Literary Sources during the 4th-1st Millennia B.C.E. Ph.D. dissertation, Hebrew University, Jerusalem. (Hebrew with English summary).
Evenari, M.; Aharoni, Y.; Shanan, L.; and Tadmor, N.H.
1958 "The Ancient Desert Agriculture of the Negev". *Israel Exploration Journal* 8:241-43.
Evenari, M.; Shanan, L.; and Tadmor N.H.
1971 *The Negev – The Challenge of a Desert*. Cambridge, MA: Harvard University.
Fritz, V., and Kempinski, A.
1983 *Ergebnisse der Ausgrabungen Hirbet 'El-Msas (Tel Masos) 1972-1975*. Wiesbaden: Harrassowitz.
Gal, Z.
1992 *The Lower Galilee in the Iron Age*. ASOR Dissertation Series 8, Winona Lake: Eisenbrauns.
Glueck, N.,
1959 *Rivers in the Desert*. London: Weidenfeld and Nicolson.
Haiman, M.
1986 *Map of Har Hamran Southwest (198)*. Jerusalem: The Archaeological Survey of Israel. Israel Antiquities Authority
1991 *Map of Mitzpe Ramon (200)*. Jerusalem: The Archaeological Survey of Israel. Israel Antiquities Authority.
1993 *Map of Nahal Aqrab (199)*. Jerusalem: The Archaeological Survey of Israel. Israel Antiquities Authority.
1998 *Map of Har Ramon (203)*. Jerusalem: The Archaeological Survey of Israel. Israel Antiquities Authority.
2003 "The 10th Century B.C. Settlement of the Negev Highlands and Iron Age Rural Palestine" in *The Rural Landscape of Ancient Israel*, eds. S. Dar; A. Maeir; and Z. Safrai. Oxford: British Archaeological Reports, International Series 1121. pp. 71-90.
In press *Survey in the Area of Ain el Gdirat and Ain Qadis*, Israel Antiquities Authority.
Lender, Y.
1990 *Map of Har Nafha (196)*. Jerusalem: Archaeological Survey of Israel. Israel Antiquities Authority.
Malamat, A.
1983 *Israel in Biblical Times, Historical Essays*. Jerusalem: Mosad Bialik (Hebrew).
Marx, E.
1974 *Bedouin of the Negev*, Kfar Habad: Reshafim (Hebrew).
Mazar, A.
1982 "Iron Age Fortresses in the Judaean Hills". *Palestine Exploration Quarterly* 114:90-103.
Mazar, B.
1957 "The Campaign of Pharaoh Shishak to Palestine". *Vetus Testamentum Supplement* IV: 57-66.
Meshel, Z.
1974 *The Negev in the Time of the Judaean Kings*. Unpublished Ph.D. dissertation, Tel Aviv University (Hebrew).
1994 "The 'Aharoni Fortress' Near Quseima and the "Israelite Fortresses" in the Negev". *Bulletin of the American Schools of Oriental Research* 294: 39-67.
Negev, A.
1979 "Water in the Desert and the Nature of the Iron Age Fortresses in the Negev Uplands". *Cathedra* 11:29-37 (Hebrew).
Rothenberg, B.
1967 *Tzfunot Negev*. Tel Aviv: Masada (Hebrew).
Shmueli, A.
1977 "Bedouin Settlements in the Desert Frontier" in *The Desert, Past-Present-Future*, ed. E. Zohar. Tel Aviv: Reshafim (Hebrew). pp. 211-216.
1980 *Nomadism About to Cease*. Tel Aviv: Dvir (Hebrew).
Woolley, C.L., and Lawrence, T.E.
1914-1915 *The Wilderness of Zin*. London: Palestine Exploration Fund Annual III.
Zertal, A.
1996 *The Manasseh Hill Country Survey II*. Haifa: University. Tel-Aviv: Ministry of Defence. (Hebrew).

NOMADS, EMPIRES AND CIVILIZATIONS: GREAT AND LITTLE TRADITIONS AND THE HISTORICAL LANDSCAPE OF THE SOUTHERN LEVANT[1]

Øystein S. LaBianca and Kristen R. Witzel

Abstract

In this paper we introduce Robert Redfield's (1962) great and little traditions framework as an aid to investigating how nomad-sedentary interactions have been constituted throughout historical times in the Southern Levant. Examples of great traditions that have, to varying degrees, shaped such interactions include the Egyptian, Mesopotamian, Arabian, Canaanite, Greek/Hellenistic, Roman, Byzantine, Islamic and Modern Capitalist. Examples of little traditions by means of which local populations have adapted to successive and often competing imperial projects and externally imposed traditions include mixed agro-pastoralism, low-risk water management, fluid shared commons, residential flexibility, tribalism, hospitality and honour. The paper ends by including several very preliminary hypotheses suggesting specific ways in which various great traditions have impacted the configuration of nomad-sedentary relations in the Southern Levant.

Introduction

Ever since Fredrik Barth's (1961) groundbreaking study of the nomadic Bassari tribesmen of Iran, anthropologists have realized that pastoral societies typically do not exist apart from agricultural societies, but as a part of them (c.f. Nelson 1973). This symbiosis of pastoral nomads and agriculturalists has also been well attested in the historical and ethnographic record of the Levant, including in Israel (Marx 1967), Jordan (Lancaster 1981) Palestine (Doumani 1995: 201-205) and Syria (Lewis 1954). More recently archaeologists and historians have also come to embrace the idea that nomads and sedentaries are part of a dynamic regional settlement system in which both play significant parts (Adams 1978; 1981, Rowton 1976; Finkelstein 1995). LaBianca's (1990) research in Jordan has involved documenting cycles of intensification and abatement in the local food system which, in turn, has been correlated with episodes of sedentarization and nomadization over time and space.

This new emphasis on viewing pastoral nomadism as enmeshed within, rather than as separate from, the wider society has important consequences for how research on the history and archaeology of pastoral nomadism in the Levant is conducted. The purpose of this paper is to highlight some of the consequences for nomadism research of this new perspective. In particular, the paper argues the case for nomad-sedentary interactions being viewed and studied as part and parcel of global geo-political events and processes involving empires and civilizations. The paper is thus one more call for a stronger link between anthropology, archaeology, epigraphy and history as a means to further advance research on long-term interactions between pastoralists and agriculturalists in the Levant (c.f. Cohn 1980; LaBianca and Scham 2005).

Limitations of the Food System Perspective

Past efforts by LaBianca (1990) to devise a framework for studying long-term interactions between nomads and sedentaries in the Madaba Plains region of Central Transjordan have centred on the food system concept and related notions such as intensification and abatement and sedentarization and nomadization. Food systems research is founded on the assumption that the quest for food security has been a primary concern of individuals, households, local communities and regional polities of various kinds since prehistoric times. While there are, no doubt, other concerns that have shaped people's lives throughout the centuries in this region, there can be little doubt that the daily routines of most ordinary residents of the Southern Levant have been impacted more profoundly by activities and traditions related to the quest for food than any other concern.

The basis for this assertion can be best grasped by considering the scope of activities and traditions involved in making food available (LaBianca 1993). These begin with procurement of food, whether by hunting or gathering, cultivation of plants, animal husbandry, or other means. They include such post-harvest activities as processing, distributing, storing, securing and preserving of food. And they include activities related to preparing, serving, consuming and disposal of food remains. When

[1] The original draft of this paper was prepared during April and May 2004 while LaBianca was a visiting researcher at the Center for Development Studies, University of Bergen. The paper was substantially revised thanks to feedback received following presentation at the 9th International Congress on the History and Archaeology of Jordan in Petra and the Albright Conference "On the Fringe of Society" in Jerusalem, both during June of 2004. Completion of the paper has been possible thanks to financial support for research and writing provided by Andrews University and by the Norwegian Research Council in connection with Global Moments Levant – a research project based at the University of Bergen in Norway.

the work and effort involved in each and all of these activities are reckoned with, it can easily be seen that activities and traditions related to the quest for food dominated the daily routines of the vast majority of men, women and children throughout history. It is only in the past century, with the rise of factory farming and fast foods, that ordinary people have been liberated from the daily demands of these routines.

While these concepts have been found to be quite useful as means to reconstruct long-term changes in the agricultural landscape and daily life routines of successive generations of occupants of the Madaba Plains, they have proven insufficient in several important respects. For example, a serious short-coming of analysis focused solely on reconstruction of food system configurations is that it tends to minimize attention to culture-historical contexts, thus leaving the impression that daily life in a rural village such as Hesban during Iron II was very similar to that during Late Roman and Mamluk times. While to a certain extent, this was true – especially with regard to how people provided for their food – we can be quite certain that different languages were spoken, that the sorts of things that could be bought in local markets and that the "news from abroad" would have been substantially different during each period.

Another problem with the food system perspective is that it does not go far enough in helping us understand why, over the centuries and millennia, peaks and valleys occur in the intensity levels of the local food system. For example, periods when the system reaches peak levels in the Hesban area are Early Bronze, Late Iron, Early Roman, Byzantine and Mamluk. Low points – valleys – occur during Middle and Late Bronze, Early Iron, Persian-Hellenistic, Late Roman, Abbasid, and Ottoman. It is due to these shortcomings of the food systems approach that the quest for a more comprehensive framework for studying nomad-sedentary interactions – a framework that takes into consideration the role of imperial projects and civilizational legacies – has gotten underway (LaBianca in press).

Civilizations, Empires and Great Traditions

To a large extent, anthropological research on 'civilizations' as a type of human social organization has focused on issues related to their origin and early evolution (Adams 1978; Lamberg-Karlovsky and Sabloff 1995). Much research has been devoted, for example, to ascertaining whether a particular civilization's origin was 'pristine' or 'secondary' – i.e. whether it evolved de-novo (and thus is pristine in its origin) or as a consequence of interaction with pre-existing civilizations or states (secondary origin). To this line of research another must be added – anthropological and archaeological studies of complex society in the Old and New World (Steward 1955; Eisenstadt 1961; Kushner 1969; Sinopoli 1994; Smith 1974; Moore 1974; Hole 1995); including explorations of the role of pastoralism within such societies (several papers in Bar-Yosef and Khazanov, eds 1992).

An especially important contribution to this second line of research is Robert Redfield's pioneering research on the 'dynamics of traditions' (1947; 1960). Redfield's work is, according to Kushner (1969:83), the "most significant contribution of social anthropology to the topic of civilization." At the core of his effort to grasp the nature of civilizations is the distinction between "great traditions" that originate from the urban epicentres of great civilizations and "little traditions" to which villagers or folk societies linked to such centres tend to cling. While in his earlier work Redfield tended to highlight the differences between these two dimensions of complex societies; in his later and more mature work, he adopts a more symbiotic view that stresses the emergent and complex interactions that regulate relationships between them (Redfield 1947; 1962: 134, 137; Kushner 1969: 83). The terminology of great and little traditions entered Middle East anthropology via Gustav von Grunebaum (1955), Redfield's colleague at the University of Chicago, who adopted it as a means to understand the interaction between universal vs. local Islam (cf Manger 1999).

The problem of how great and little traditions interact became a major area of research for McKim Marriott (1955), a student of Redfield's. According to Marriott "a great tradition owes its existence to two processes. The primary process is *universalization,* by which Marriott meant the carrying forward of the material which is already present in the little traditions in the villages to a body which 'universalized' the knowledge into a great tradition. The second process is *parochalization* or the 'downward spread' to the villages of the great tradition. Both universalization and parochalization are characterized by transformations, and there are gaps in communication which the communities fill at their own discretion" (Marriott 1955 as cited in Odner 2000: 34).

In the writings of Redfield and others who have adopted his structure of traditions terminology (von Grunebaum 1955; Bharati 1978; Bodley 2004) the concepts of 'civilization' and 'great tradition' are closely related terms and at times are used interchangeably. In von Grunebaum's (1955) work on Islamic civilization, for example, the great tradition is universal Islam; and in Bodley's (2004) widely used *Cultural Anthropology* textbook, the civilizations of the Hindu, Islamic and Chinese are all referred to as 'great traditions.'[2]

For our purposes here, a definition of the concept of civilization will be proposed that highlights the enduring legacy of civilizations. To this end concepts from

[2] Bodley (2004:229) defines "great tradition" as "the culture of the elite in a state-organized society with a written tradition that is not fully shared by non-literate, village-level commoners."

astronomy[3] – luminous, radiant and constellation – are employed as follows: *a 'civilization' is a luminous constellation of radiant attitudes, beliefs, institutions, values, elite cultural traditions and works of art, artisanry and architecture that emanate from a particular urban epicentre or centres such as ancient Sumer, Thebes, Athens or Rome.*

Civilizations are luminous when they emit "light" that advances the development of human societies and cultures. Such light might be contributions to the development of writing and language or innovations in such varied fields as religion, technology, commerce, science or the arts.

Civilizations are radiant in two senses. First, they typically radiate from some epicentre, usually one or more great urban capitals within their heartlands. Examples include ancient Sumer in the case of Mesopotamian civilization; Thebes in the case of Egyptian Civilization; and Athens in the case of Greek/Hellenistic civilization. Second, civilizations are radiant in the sense that their luminosity continues long after their "golden age." Thus, the contributions of the ancient Greeks to the development of philosophy and natural science continued to radiate well beyond the middle of the first millennium B.C. – the Greek Golden Age – as evidenced by Greek influence on subsequent civilizations such as those of the Romans, Byzantines, and Medieval Christians.

Civilizations resemble constellations because they are distinguishable by means of an enduring configuration of component parts including attitudes, beliefs, values, elite cultural traditions and works of art, artisanship and architecture. These configurations are what we commonly recognize as Mesopotamian civilization, Egyptian civilization, Roman civilization, Islamic civilization and so on.

Significantly, however, these configurations are always somewhat varied and localized. Very luminous and radiant civilizations may metamorphose through successive distinguishable phases. A good example is Egyptian civilization, which lasted over three millennia starting with the pre-dynastic phase, and continuing through the Early, Middle, and Late Dynastic.

A civilization's "great traditions" are normative principles and behaviours propagated by its literate elites who are the primary custodians of the canonical texts. Its little traditions, on the other hand, are the conglomerate of vernacular or local knowledge and practices not derived from a canonical text, but considered normative by the largely non-literate masses. As already indicated, 'great traditions' are disseminated by means of universalizing agents. Such agents include empires, dynasties, and sometimes religious movements. Agents disseminate 'great traditions' by means of projects – concrete undertakings of various kinds that expand the reach of a 'great tradition.'

For our purposes here we shall define a 'great tradition' as *universalized collective heritage and knowledge.* This is a rather simple and straightforward definition that emphasizes the most characteristic aspect of great traditions – namely there being traditions that have been universalized by the aid of some dynastic or imperial agency.

Geographical Considerations

As a background for understanding the role of great and little traditions in shaping the historical landscape of the Southern Levant, four notable geographical characteristics of the region must be reckoned with:

First is the region's geographic position astride an intercontinental land bridge linking together the continents of Africa, Europe and Asia (Levy 1998). Having served since prehistoric times as a vital corridor of communication, migration, and trade, the Southern Levant has long been a coveted piece of real estate over which rival dynasties in Egypt, Mesopotamia, Anatolia, Greece, Rome and Western Europe have sought to exert control and domination.

Second is its proximity to the Arabian steppe. For multiple millennia, this steppe has served not only as the desert headquarters of long-distance caravan trade, but also as a wellspring of Bedouin culture and aspirations. Since earliest antiquity Bedouin tribes from the Arabian steppe have infiltrated the fertile highlands of both Transjordan and Cis-Jordan, replenishing its population while emboldening it in its resistance to foreign domination and control (Eph'al 1982).

Third is the Mediterranean Sea (Braudel 1972; Horden and Purcell 2000) that connects the Southern Levant to ports of call all around the Mediterranean and beyond. Coastal cities such as Tyre and Sidon, Ashkelon and Caesarea are examples of harbours through which trade goods could be channelled from inland cities and towns to distant ports of call and vice versa. At various points in time this same coastline has also served as a point of access to the region by seafaring invaders such as the Philistines, the Greeks, the Romans, and the Crusaders.

And fourth is its natural endowment of which availability of water is the single most important factor impacting human livelihoods (Hopkins 1985; Lowi 1995). As a general rule, rainfall is most plentiful in highland regions, especially as one moves northward and westward across Transjordan and the Southern Levant. Sedentary agriculture is thus most sustainable in the well-watered highlands of both sides of the Rift Valley and along the

[3] We are not alone in borrowing images from astronomy to describe the nature of civilization. In the introduction to *A History of Civilization* Fernand Braudel (1993: 14) illustrates the process by means of which a specific cultural trait is transmitted through time and space from a vanished civilization by comparing it to a star whose "light still reaches us."

slopes of river valleys and wadies that drain eastward and westward from these parallel highlands. In these drainages natural springs and man-made cisterns, dams and terraces make water available year-around in this otherwise semi-arid landscape. These four factors combine to make the Southern Levant cross-roads of civilizations and a point of often violent conjuncture of great and little traditions.

Great Traditions of the Southern Levant

Below we propose eight great traditions or civilizations that arguably have contributed to a significant degree to producing the archaeological and historical landscape of the Southern Levant over the past five millennia. These eight traditions represent at best a first approximation to give an idea of what such a line-up might look like, and we fully expect that as research on imperial powers and civilizational legacies impacting the Levant continues, there will be significant modifications to this line-up. Since the salient features of each have been discussed in somewhat greater detail elsewhere (LaBianca in press) we resort here to a bullet outline format as a means to highlight a few main points. To these points, most of which reflect the authors' perspective as food systems researchers, others could be added representing other disciplines and points of view than those of the writers. Hopefully this will be done in due course. In the meantime, the present line-up will have to do and will serve as a starting point for future great traditions research in the Levant. Most of the great traditions named should not come as any surprise to scholars working in the region.[4]

The Egyptian Great Tradition[5]

- Developed in the Nile Basin in response to the annual gentle flooding of the Nile River
- Hypothesized core components include: 1) production of wheat, barley, pulse and flax by village farmers in naturally irrigated flood basins; 2) centralized regulation of the agricultural calendar by a religious elite with the pharaoh at the top; 3) diversion of agricultural labour to public works such as pyramid construction during flooding season; 4) on and off control of provinces or *nomes* by central governments and 5) reliance on hieroglyphics as a means to codify knowledge and practices essential to the maintenance of the tradition.

The Mesopotamian Great Tradition[6]

- Developed in Southern Mesopotamia in response to annual – often violent – late spring flooding of the Euphrates and especially the Tigris rivers
- Hypothesized core elements include: 1) irrigation of cereal fields and pasturelands by means of networks of man-made canals transporting water from the main channels of these two rivers; 2) application of various social technologies – for example, priestly regulation of the agricultural calendar – as a means to ensure elite control of agricultural production; 3) protection and centralized control of agricultural villages and canals by means of city-states and 4) reliance on cuneiform script as a means to codify knowledge and practices essential to the maintenance of the tradition.

The Arabian Great Tradition[7]

- Developed in the desert oasis and arid steppe of the Arabian Peninsula
- Hypothesized core components include: 1) adaptation to arid desert conditions by means of migratory production of sheep and goats; 2) symbiotic, often predatory, interactions with agricultural populations along the Fertile Crescent; 3) reliance on kin-based or tribal social organization as a means to protect rangelands, watering places and households and 4) reliance on a group of Semitic language and scripts, such as Thamudic, Safaitic, and Old South Arabic, as a means to codify knowledge and practices essential to the maintenance of the tradition.

The Canaanite Great Tradition[8]

- Developed in response to the ecological conditions along the Eastern Mediterranean, in particular the fertile highlands of coastal and southern inland Syria (including Lebanon), Israel, Palestine and Jordan.
- Hypothesized core components include: 1) local-level collection of rainwater on fertile slopes by means of terracing, diversion dams and cisterns; 2) agriculture emphasizing dry-farming of cereals and terrace-based production of olives and grapes; 3) significant capacity for sedentarization and nomadization as a means to manage risks and respond to new opportunities and threats reliance on extended family networks, a temple-based religious cult, and at times tribal kings, as a means to organize production and protect against enemies

[4] Surely some will wonder why I have not included in this line-up a mention of a Jewish Great Tradition. The main reason for not including it here is that, as Eisenstadt (2005) has noted, it is a de-territorialized civilization—which makes it a very special case. I welcome debate and suggestions on how best to incorporate the enormous contribution of Jews into this line-up of great traditions.

[5] See Butzer 1976, Eyre 1995: 186; Leprohon 1995: 273-287; Manley 1996; Mendelssohn 1974; Robins 1995: 1811-1812; Wetterstrom and Murry 2001.

[6] See Adams 1978, 1981; Eyre 1995; Postgate 1990; Wittfogel 1957; Lamberg-Karlovsky and Sabloff 1995:139-181; Also see Woolley 1965; Kramer 1971.

[7] See Eph'al 1982:12-17; Guillaume 1954: 2-5; von Soden 1994: 23-24; and Zurayk 1949. Also see Hitti 2002 and Lewis 1992;

[8] See Ahlstrom 1995: 587-631; Gragg 1997: 516-527; Gray 1964; Hopkins 1985; LaBianca and Younker 1995; LaBianca 1997; Tubb 1998:13. Also see Albright 1949; Redford 1992.

and 4) reliance on a specific Northwest Semitic language and script (Amorite, Ammonite, Edomite, Hebrew, Moabite) as a means to codify knowledge and practices essential to the maintenance of the tradition

The Greek/Hellenistic Great Tradition[9]

- Resembles to some degree that of the Canaanite Great Tradition.
- Hypothesized core components include: 1) local-level collection of rainwater in fertile valleys and slopes by means of terracing, diversion dams and cisterns; 2) production of cereals by means of dry farming on open plains and orchards planted to olives, figs and grapes on terraced hillsides; 3) transhumant production of sheep and goats on mountain pastures and stubble fields in cultivated areas; 4) reliance on extended family networks as a means to organize production and protect against droughts and enemies; and 5) use of the Greek language and script as a means to codify knowledge and practices essential to the maintenance of the tradition

The Roman Great Tradition[10]

- Is in many ways a continuation of the Greek/Hellenistic Great Tradition, but can be distinguished from this earlier tradition because of its heavy emphasis on maximization of agricultural production and yield.
- Hypothesized core components include 1) maximization of water supply through addition of aqueducts, waterwheels, pumps, large underground cisterns and reservoirs for storing water, and drainage systems for managing runoff; 2) maximization of land area available for agricultural production through removal of forests and draining of swamps; 3) maximization of yield through application of green manure, fertilizers, and rotation of crops; 4) maximization of crop and stock yields through breeding and improved husbandry practices; 5) maximization of farm labour through increased use of slaves and hired personnel and 6) use of the Latin language and script as a means to codify knowledge and practices essential to the maintenance of the tradition.

The Byzantine Great Tradition[11]

- Continued to a significant extent the emphasis on maximization introduced by the Romans, but with modifications attributable in great measure to the rise of Byzantine Christianity, centred in Constantinople.
- Hypothesized core components include 1) continuation, and in certain locations, expansion of Roman system for maximizing water supply; 2) addition of monasteries and estate farms as centres of agricultural production and distribution; 3) intensification of cash crop production of olives and grapes; 4) increased hierarchical organization of production due to increased concentration of political power in the hands of bishops at the expense of city endowments and private citizens and 5) use of the Greek language and script as a means to codify essential knowledge and practices. The emperor was at the apex of Byzantine Christendom, as head of both the church and the state – a polity referred to by some scholars as caesaropapism (Margoulias 1970:1-16).

The Islamic Great Tradition[12]

- Has roots in the Arabian Great Tradition.
- Was founded in the 6th century A.D. by the prophet Muhammad, and incubated among caravaneers and desert tribesmen of the Saudi Arabian desert. From its beginnings in Mecca and Medina, Islam spread rapidly to the great population centres of Late antiquity – to Baghdad, Cairo, Damascus and Jerusalem – to become the dominant religion of the Arabs throughout all the lands of the Middle East and far beyond.
- Hypothesized core components include 1) irrigation agriculture in the great river basins of Egypt and Mesopotamia; 2) dry-farming of cereals in the mountainous regions of the Levant and Northern Syria; 3) pastoral nomadism in the deserts of Arabia and North Africa; 4) extensive inter-regional trade and exchange in foodstuffs and other goods facilitated by caravaneers and universal Islam; 5) a world view and elite culture anchored in the teachings of the Koran and the Prophet Muhammad and 6) use of the Arabic language and script as a means to codify knowledge and practices essential to the maintenance of the tradition.

The Modern Capitalist Great Tradition[13]

- Born of the "great transformations" which took place in Europe during the 16th-18th centuries – including the Renaissance, Reformation, Enlightenment and the scientific and industrial revolutions – the Modern Capitalist Great Tradition has had enormous consequences for how

---

[9] See Gragg 1997:516-527; Hughes 1975: 68-82. Also see Hamilton 1993a; Freeman 1999; Boardmann *et al.* 1986; Sansone 2004.

[10] See Foss 2002; Hamilton 1993b: 129-130; Hughes 1975; and White 1970: 86-383. Also see Luttwak1976; Christ 1984; Boatwright *et al.* 2004; Boardmann *et al.* 1988; Butcher 2003; and Sartre 2005.

[11] See Evenari *et al* 1971; Foss 2002:71-95; Margoulias 1970:1-16; Patrich 1995: 483; Reifenberg 1955. Also see Shahid 1984a; 1984b; 1989; 1995; Kaegi 1992; Laiou and Maguire 1992; Cavallo 1997; Veyne 1987; Mango 2002; El Cheikh 2004.

[12] See Guillaume 1954; Tritton 1951: 9-22, Walker 1999. Also see Crone 1980 and Kennedy 2004.

[13] See Wallerstein 1976; Bodley 2000: 308-309; Braudel 1995; Friedman 1998; and Wolf 1997. Also see Castells 1996, 1997, 1998.

food is produced, stored, distributed, prepared and consumed around the world. What sets this new tradition apart is its reliance on fossil fuels, fertilizers, mechanization, and international markets.

- Under this new system, food is produced by means of factory farming methods for the sake of profit
- As was the case with the Roman great tradition, the goal is maximization, but this time not only by means of application of technology, but also by expansion of demand through globalization.

Little Traditions of the Southern Levant

Elsewhere LaBianca (forthcoming) has defined "little traditions" as *localized indigenous heritage and knowledge.* What sets little traditions apart from great traditions is their local character and autonomy. They represent, in other words, knowledge and practices that have been passed on from generation to generation without being dependent on the aid of dynastic or imperial universalizing agents for their transmission. They help explain how individual households and whole tribes for centuries and millennia have been able to shift back and forth between sedentary and nomadic livelihoods in response to changing environmental and socio-political circumstances.[14] Seven such little traditions have come to light as a result of research on ancient food systems at Tell Hesban in Jordan. A summary of what these are is outlined below:

Mixed agro-pastoralism[15]

- Refers generally to the practice of combining crop cultivation with animal husbandry. Although a wide range of "mixes" are possible, the classic pattern in the Eastern Mediterranean is cultivation of cereals and small stock, especially sheep and goats.
- The advantage of this practice is that after the fields have been harvested, the small stock can graze on the stubble while dropping fertilizing dung on the ground.
- Where climate, economic, and security conditions permit, this practice may be combined with production of legumes or the cultivation of olives and grapes on slopes and terraces.

Transhumance[16]

- Involves people moving with their mixed herds between lowland and highland regions on an annual basis in order to maximize productivity of both lands and herds.
- Transhumance allows people to respond to different economic, political or environmental opportunities or threats by adjusting the mix in terms of emphasis: cereals, legumes, olives, grapes, other fruits, sheep, goats, pigs or poultry.

Low-risk water management[17]

- Refers to the practice of relying on natural springs and man-made cisterns for water to meet household and farming needs.
- Compared with large scale systems for collecting, transporting and storing water, such as dams, aqueducts, and reservoirs, which are difficult to construct, require a great deal of labour to maintain, and are vulnerable targets in times of conflict, natural springs and cisterns represent a comparably low-risk approach to water management.

Fluid shared commons[18]

- In this system, pastures and croplands do not belong exclusively to one family or household, but are used by a range of different families representing one or more communities or tribes.
- Over a given period of time, the territory of such shared commons may shift as individual families or communities adjust their production strategy to emphasize herding or agriculture; thus the boundaries of shared commons are typically fluid.

Residential flexibility[19]

- This tradition allows families to shift the location of their production activities between houses, and caves or tents.
- Shifts can occur at several levels. In some cases, a given household may split up, with some members moving into tents to be closer to their herds, and others staying behind in the house to look after the crops. In other cases, households may live in tents during the warm months of the year, and move into caves or houses during the colder months.

Tribalism[20]

- Tribalism can be defined as a flexible polity

[14] Elsewhere (LaBianca 1997) I have referred to these little traditions as "indigenous hardiness structures" in order to emphasize their origin among local residents (indigenous); their function in fortifying against hunger and famine (hardiness); and their persistence as cultural practices (structures).

[15] See Burckhardt 1831: 1; Lewis 1987: 3; Wåhlin 1997.

[16] See Libbey and Hoskins 1905: 252-253; Lewis 1987: 3-5; Condor 1892: 129; Burckhardt 1831: 227-228.

[17] See Burckhardt 1831: 228,229; Forder 1909: 236-237; Tristram 1873: 119, 322; Libbey and Hoskins, 1905: 173.

[18] See Burckhardt 1831 (Vol. 1): 14, 247; Layne 1994: 48 (note 5); Lewis 1987: 23.

[19] See Forder 1909: 56,123 ; Tristram 1873: 189, 248 ; Glubb 1978: 37.

[20] See Dickson 1959: 52; Lancaster and Lancaster 1992; Lancaster and Lancaster 1999: 12; Nugent and
Sanchez, 1994: 94-95, Layne 1994: 41; Forder 1909: 3; Glubb 1978: 26, 65-68, 84.

involving strong in-group loyalty based on variously fluid notions of common lineal descent.

- This highly flexible system of local level political organization has provided indigenous residents of the Southern Levant with a means by which small groups of kin have been able to adjust consistently and successfully to political and economic domination and change.

Hospitality[21]

- Is a means of transmitting vital information within a tribal community, and between members of a tribe and outsiders.
- This institution plays a crucial practical role in maintaining solidarity between members of the tribe, and in facilitating the transmission of information vital to group solidarity and survival.

Honour[22]

- Provides a method of social control whereby tribes have been able to police themselves without actually having to rely on a paid police force or a codified system of civil and/or criminal law.
- Is a critical component of maintaining social order without having to rely on externally imposed systems of law and punishment.
- Provides an incentive for economic success necessary to maintain ideals of hospitality and generosity.[23]

Great and Little Traditions and the Historical Landscape of the Levant

The great and little traditions outlined above, when combined with the four geographical considerations outlined earlier, go a long way toward accounting for both temporal and regional variation in the historical landscape of the Southern Levant. They represent, I would argue, the key factors responsible for producing changes over time and space in the extent to which a particular region is dominated by farmers or shepherds – by towns and villages or nomads. They also account for the extent to which, in a given time and place, sedentarization or nomadization is in ascendance or decadence respectively.

While much has been learned about how particular great traditions have facilitated the process of sedentarization in the Southern Levant (LaBianca 1990); and quite a lot has also been learned about how little traditions facilitate movement back and forth between nomadic and sedentary ways (LaBianca 1997); very little is understood about how particular great traditions have contributed to the rise and persistence of nomadism. An important question, in this regard, is whether in fact certain great traditions have actively fostered nomadization and nomadism or whether the latter are primarily the result of disinvestment and neglect by a given imperial or dynastic agency. It is nevertheless possible to posit a few hypotheses linking such agency action with the rise and persistence of nomadism in the Levant.

To begin with, let us take a look at the Mesopotamian Great Tradition. An influential perspective, in this regard, is that of Rowton (1976), a Mesopotamian scholar, who has argued that throughout the Semitic world a dimorphic social structure existed – one emphasizing urban sedentism, the other emphasizing pastoral nomadism. Between these two a relationship of symbiosis existed, but, in his view, the two groups were separate, not really a part of the same social system. I believe this was much less the case in the Southern Levant, as here there really were no great urban centres. Instead, the predominant way of life was rural, and therefore, to a much greater extent than in Mesopotamia, and Egypt for that matter, the lives of agriculturalists and nomads were very significantly enmeshed. Thus, the Old Testament portrays Abraham as being a nomad, but his brother-in-law, Lot, is an urban dweller. To the extent that some scholars (Dever 1998) have noted the existence of a "nomadic ideal" in the Old Testament, it no doubt reflects the strong influence of the Arabian Great Tradition on the region.

Unlike the Mesopotamian and Arabian great traditions, the Hellenistic, Roman and Byzantine great traditions were all much more favourable toward sedentary modes of existence, and actively opposed nomadic ones. A consequence of this was that during Greco-Roman and Byzantine times, nomadism emerged among the Arabs as a weapon of resistance. Those who refused to cooperate with their Hellenistic and Roman masters would join federations of nomads in the Arabian Steppe, and launch periodic attacks on their Hellenized brothers and sisters living under Roman occupation (Shahid 1984a; 1984b; 1989; 1995).

To a very significant extent, Muhammad was able to benefit from this build-up in the Arabian Steppe of disenfranchised Arab tribes. Under the banner of Islam – a religion emphasizing the equality of all men – he set out

[21] See Burckhardt 1831: 341,342; Condor 1892: 129; Forder 1909: 156-157, 159,
263-264; Glubb 1978: 156-158.

[22] See Burckhardt 1831: 115,134; Condor 1892: 322-324, 352; Forder 1909: 29-30.

[23] Burckhardt (1831) writes, "A sheikh does not derive any yearly income from his tribe, or camp; on the contrary, he is obliged to support his title by considerable disbursements, and to extend his influence by great liberality. It is expected that he should treat strangers in a better style than any other person of the tribe; that he should maintain the poor, an divide among his friends whatever presents he may receive. His means of defraying these expenses are, the tribute which he extracts from the Syrian villages, and his emoluments from the Mekka pilgrim caravan.... A living sheikh is sometimes deposed, and a more generous man elected in his place. (pp. 118, 119)". See also Burckhardt 1831: 248, 284-285.

to dismantle the systemic injustices that were rampant in areas under Greco-Roman and Christian control. Under the new Islamic Great Tradition the nomadic ideal was again restored, with the consequence that the separation that had occurred between nomads and sedentaries under the former regime was replaced by a renewed symbiosis of the two modes of existence.

While the full impact of the Modern Capitalist Tradition on pastoral nomadism is still not known, there is some evidence that the "profit motive" has begun to turn both nomads and farmers into agribusiness men and women. Animals and crops are thus no longer being raised on a subsistence basis, but rather as a source of cash income. Mechanization is aiding this process, in that nomads turn to trucks to transport their herds, and settled farmers mount tractors and combines to produce their harvest. It is doubtful that mixed farming will survive here any more than it has in other parts of the world that have come under the domination of the Modern Capitalist System.

Future Research

What we have attempted to do in this paper is to lay open for discussion and criticism a possible programme of study for advancing research on the impact of imperial powers and civilizational legacies on nomad-sedentary interactions in the Levant. To this end, we have employed the framework of great and little traditions as a means to organize and characterize the various structural elements that must be reckoned with in attempting an anthropological history of such interactions in the region. The line-up of salient great traditions that we have proposed – although no doubt not to everyone's liking – is common knowledge among scholars working in the region. The little traditions are common knowledge among the rural masses of the region, although not necessarily among scholars.

What has been posited here is a framework for conceptualizing the various strands of influences that have cooperated over the long-term to shape the historical landscape of the Southern Levant over time. Much more research is needed to clarify the salient features of each of the great and little traditions highlighted and also to determine which other traditions should be added. Future research is also needed to understand the synergistic interactions of little traditions with imperial projects and civilizational legacies in producing particular historical landscapes and societies throughout the region. To this end anthropological, archaeological and historical research must proceed hand in hand so that the full extent of regional variation and historical particularity that has existed in the region will come to light.

Bibliography[24]

Adams, R. McC.
1978 "Strategies of Maximization, Stability, and Resilience in Mesopotamian Society, Settlement, and Agriculture". *Proceedings of the American Philosophical Society* 122/5: 329-35.
1981 *Heartland of Cities: Surveys of Ancient Settlement and Land Use on the Central Floodplain of the Euphrates.* Chicago: University of Chicago.

Ahlström, G. W.
1995 "Administration of the State in Canaan and Ancient Israel" in *Civilizations of the Ancient Near East,* vols. 3 and 4, eds J. Sasson; J. Baines; G. Beckman; and K. Rubinson. Peabody, MA: Hendrickson. pp. 587-604.

Albright, W. F.
1949 *The Archaeology of Palestine.* London: Penguin.

Barth, F.
1961 *Nomads of South Persia: The Basseri Tribe of the Khamseh Confederacy.* Boston: The Little, Brown Series in Anthropology.

Bar-Yosef, O., and Khazanov, A. eds
1992 *Pastoralism in the Levant: Archaeological Materials in Anthropological Perspective.* Monographs in World Archaeology No. 10. Madison, Wisconsin: Prehistory.

Bharati, S.A.
1978 *Great Tradition and Little Tradition: Indological Investigations in Cultural Anthropology.* Varanasi, India: Chowkhamba Sanskrit Series Vol. XCVI.

Boardman, J.; Griffin, J.; and Murray, O., eds
1986 *The Oxford Illustrated History of Greece and the Hellenistic World.* Oxford: Oxford University.
1988 *The Oxford Illustrated History of the Roman World.* Oxford: Oxford University.

Boatwright, M.; Gargola, D.; and Talbert, R.J.A.
2004 *The Romans: From Village to Empire.* Oxford: Oxford University.

Bodley, J.
2004 *Cultural Anthropology: Tribes, States, and the Global System.* 3rd Edition. Mountain View, CA.: Mayfield.

Braudel, F.
1972 *The Mediterranean and the Mediterranean World in the Age of Phillip II.* 2 Volumes. New York: Harpers and Row.
1993 *A History of Civilizations.* London: Penguin.
1995 *A History of Civilizations.* New York: Penguin.

[24] This list of references is quite long as I have chosen to include a number of classic or recently published volumes dealing with the various great traditions. The selection is by no means exhaustive nor especially authoritative, but rather, it gives an indication of the sorts of secondary sources that are available today to archaeologists interested in gaining a deeper understanding of the various great traditions.

Burckhardt, J.L.
1831 *Notes on the Bedouins and the Wahabys*, Vols. 1 and 2. London: Henry Colburn and Richard Bentley.
Butcher, K.W.
2003 *Roman Syria and the Near East*. Los Angeles: Getty.
Butzer, K.W.
1976 *Early Hydraulic Civilization in Egypt: A Study in Cultural Ecology*. Chicago: University of Chicago.
Castells, M.
1996 *The Information Age: Economy, Society and Culture. Volume 1: The Rise of the Network Society*. Oxford, and Malden, MA: Blackwell.
1997 *The Information Age: Economy, Society and Culture. Volume 2: The Power of Identity*. Oxford, and Malden, MA: Blackwell.
1998 *The Information Age: Economy, Society and Culture.Volume 3: End of Millennium*. Oxford, and Malden, MA: Blackwell
Cavallo, G., ed.
1997 *The Byzantines*. Chicago: University of Chicago.
Christ, K.
1984 *The Romans*. Berkeley: University of California.
Cohn, B. S.
1980 "History and Anthropology: The State of Play". *Comparative Studies in Society and History*, 22 No. 2.: 198-221.
Condor, C.R.
1892 *Heath and Moab: Explorations in Syria in 1881 and 1882*. London: Watt.
Crone, P.
1980 *Slaves on Horses: The Evolution of the Islamic Polity*. Cambridge: Press Syndicate of the University of Cambridge.
Dever, W.G.
1998 "Israelite Origins and the 'Nomadic Ideal': Can Archaeology Separate Fact from Fiction?" in *Mediterranean Peoples in Transition: Thirteenth to Early Tenth Centuries BCE*, eds S. Gitin; A. Mazar; and E. Stern. Jerusalem: Israel Exploration Society. pp 220-37.
Dickson, H.R.P.
1959 *The Arab of the Desert: A Glimpse into Bedawin Life in Kuwait and Saudi Arabia*. London: George Allen and Unwin.
Doumani, B.
1995 *Rediscovering Palestine: Merchants and Peasants in Jabal Nablus, 1700-1900*. Berkeley: University of California.
Eisenstadt, S.N.
1961 "Anthropological Studies of Complex Societies". *Current Anthropology* 2/3: 201-22.
2005 "The Jewish Historical Experience: Heterodox Tendencies and Political Dynamics in a De-territorialized Axial Civilization" in *Axial Civilizations and World History*, eds J. P. Arnason; S.N. Eisenstadt; and B.Wittrock. Leiden: Brill. pp. 225-51.
El Cheikh, N. M.
2004 *Byzantium Viewed by the Arabs*. Cambridge: Harvard University.
Eph'al, I.
1982 *The Ancient Arabs: Nomads on the Borders of the Fertile Crescent 9th-5th Centuries B.C.* Leiden: Brill.
Evenari, M.; Shanan, L.; and Tadmor, N.
1971 *The Negev: The Challenge of Desert*. Cambridge: Harvard University.
Eyre, C. J.
1995 "The Agricultural Cycle, Farming, and Water Management in the Ancient Near East" in *Civilizations of the Ancient Near East*, vols. 3 and 4, eds J. Sasson; J. Baines; G. Beckman; and K. Rubinson. Peabody, MA: Hendrickson. pp 175-89.
Finkelstein, I.
1995 *Living on the Fringe: The Archaeology and History of the Negev, Sinai and Neighbouring Regions in the Bronze and Iron Ages*. Sheffield: Sheffield Academic Press.
Forder, A.
1909 *Ventures Among the Arabs*. New York: Gospel Publishing House.
Foss, C.
2002 "Life in City and Country" in *The Oxford History of Byzantium*, ed. C. A. Mango. Oxford: Oxford University. pp. 71-95.
Freeman, C.
1999 *The Greek Achievement: The Foundations of the Western World*. London: Penguin.
Friedman, Th. L.
1998 *The Lexus and the Olive Tree*. New York: Anchor.
Glubb, J.B.
1978 *Arabian Adventures*. London: Cassell.
Gragg, G. B.
1997 "Semitic Languages" in *The Oxford Encyclopaedia of Archaeology in the Near East*, vol. 4, ed. E. M. Meyers. New York: Oxford University. pp 516-26.
Gray, J.
1964 *The Canaanites*. New York: Praeger.
Guillaume, A.
1954 *Islam*. London: Penguin Books.
Hamilton, E.
1993a *The Greek Way*. London: Norton.
1993b *The Roman Way*. London: Norton.
Hitti, P. K.
2002 *History of the Arabs*. New York: Palgrave McMillan.
Hole, F. A.
1995 "Assessing the Past Through Anthropological Archaeology" in *Civilizations of the Ancient Near East*, ed. J. M. Sasson. New York: Scribner. pp. 2715-27.
Hopkins, D. C.
1985 *The Highlands of Canaan: Agricultural Life in the Early Iron Age*. London: Dotesios.
Horden, P., and Purcell, N.
2000 *The Corrupting Sea: A Study of Mediterranean History*, Oxford: Blackwell.

Hughes, J. D.

1975 *Ecology in Ancient Civilizations.* Albuquerque, NM: University of New Mexico.

Kaegi, W. E.

1992 *Byzantium and the Early Islamic Conquests.* Cambridge: Cambridge University.

Kennedy, H.

2004 *When Baghdad Ruled the Muslim World: The Rise and Fall of Islam's Greatest Dynasty.* Cambridge, MS: Perseus.

Kramer, S. N.

1971 *The Sumarians: Their History, Culture and Character.* Chicago: University of Chicago.

Kushner, G.

1969 "The Anthropology of Complex Societies". *Biennial Review of Anthropology* 6:80-131.

LaBianca, Ø. S.

1990 *Hesban 1: Sedentarization and Nomadization: Food System Cycles at Hesban and Vicinity in Transjordan.* Berrien Springs, MI: Andrews University.

1993 "The fluidity of social groups in central Transjordan" in *The Middle East: Unity and Diversity*, eds H. Palva and K. S. Vikør. Copenhagen: Nordic Proceedings in Asian Studies, No. 5. NIAS. pp. 206-15.

1997 "Indigenous hardiness structures and state formation in Jordan: Towards a history Jordan's resident Arab population" in *Ethnic Encounter and Culture Change*, eds M. Sabour and K. S. Vikør. Bergen, Norway: Nordic Research on the Middle East 3. pp. 143- 57.

(in press) "Great and Little Traditions: A Framework for Studying Cultural Interaction through the Ages in Jordan" in *Studies in the History and Archaeology of Jordan No. 9.* Amman: Department of Antiquities.

LaBianca, Ø. S., and Younker, R. W.

1995 "The Kingdoms of Ammon, Moab, and Edom: The Archaeology of Society in Late Bronze/Iron Age Transjordan (CA. 1400-500 BCE)" in *The Archaeology of Society in the Holy Land*, ed. T. Levy. London: Leicester University. pp. 399-415.

LaBianca, Ø. S., and S. A. Scham

2005 *Connectivity in Antiquity: Globalization as a Long Term Historical Process*, London: Equinox.

Laiou, A. E., and Maguire, H., eds

1992 *Byzantium: A World Civilization.* Washington D.C.: Dumbarton Oaks.

Lamberg-Karlovsky, C.C., and Sabloff, J. A.

1995 *Ancient Civilizations: The Near East and Mesoamerica.* Second Edition. Prospect Heights, Ill: Waveland.

Lancaster, W.

1981 *The Rwala Bedouin Today.* Cambridge: Cambridge University.

Lancaster, W., and Lancaster, F.

1992 "Tribal Formations in the Arabian Peninsula". *Arabian Archaeology Epigraphy* 3 (2): 145-72.

1999 *People, Land and Water in the Arab Middle East: Environments and Landscapes in the Bilad ash-Sham.* Amsterdam, Harwood Academic Publishers.

Layne, L.

1994 *Home and Homeland: The Dialogics of Tribal and National Identities in Jordan.* New Jersey, Princeton University.

Leprohon, R. J.

1995 "Royal Ideology and State Administration in Pharaonic Egypt" in *Civilizations of the Ancient Near East,* vols. 3 and 4, eds J. Sasson; J. Baines; G. Beckman; and K. Rubinson. Peabody, MA: Hendrickson. pp 273-87.

Levy, Th., ed.

1998 *The Archaeology of Society in the Holy Land.* London: Leicester University.

Lewis,B.

1992 *The Arabs in History.* Oxford: Oxford University.

Lewis, N.

1954 "The Frontier of Settlement in Syria, 1800-1950". *International Affairs* (January): 48-60.

1987 *Nomads and Settlers in Syria and Jordan, 1800-1980,* Cambridge: Cambridge University.

Libbey, W., and Hoskins, F.E.

1905 *The Jordan Valley and Petra.* New York: Putnam's Sons.

Lowi, M. R.

1995 *Water and Power: The Politics of a Scarce Resource in the Jordan River Basin.* Cambridge: Cambridge University.

Luttwak, E. N.

1976 *The Grand Strategy of the Roman Empire: From the First Century A.D. to the Third.* Baltimore: Johns Hopkins University.

Manger, L., ed.

1999 *Muslim Diverstiy: Local Islam in Global Contexts.* Surrey: Curzon.

Mango, C., ed.

2002 *The Oxford History of Byzantium.* Oxford: Oxford University.

Manley, B.

1996 *The Penguin Historical Atlas of Ancient Egypt.* Great Britain: The Bath Press.

Margoulias, H. J.

1970 *Byzantine Christianity: Emperor, Church and the West.* Chicago. Rand McNally.

Marriott, M.

1955 "Little Communities in an indigenous civilization" in *Village India. Studies in the Little Community*, ed. M. Marriott. Chicago: University of Chicago. pp. 171-222.

Marx, E.

1967 *Bedouin of the Negev.* Manchester: Manchester University.

Mendelssohn, K.

1974 *The Riddle of the Pyramids.* New York: Praeger.

Moore, C. B.

1974 *Reconstructing Complex Societies.* Supplement to the Bulletin of the American Schools of Oriental Research No. 20.

Nelson, C., ed.
1973 *The Desert and the Sown: Nomads in the Wider Society.* Berkeley: University of California Institute of International Studies.
Nugent, J. B. and Sanchez, N.
1994 "Tribes, Chiefs, and Transhumance: A Comparative Analysis". *Economic Development and Cultural Change* 42,1: 87-113.
Odner, K.
2000 *Tradition and Transmission: Bantu, Indo-European, and Circumpolar Great Traditions.* Bergen Studies in Social Anthropology No. 54. Bergen: Norse Publications.
Patrich, J.
1995 "Church, State and the Transformation of Palestine – The Byzantine Period (324-640 CE)" in *The Archaeology of Society in the Holy Land,* ed. T. Levy. London: Leicester University. pp 470-87.
Postgate, J.N.
1990 "Irrigation and Cultivation in Mesopotamia Part 1, 2". *Bulletin on Sumerian Agriculture* vol. V. Cambridge: Aris & Phillips.
Redfield, M., ed.
1962 *Human Nature and the Study of Society: The Papers of Robert Redfield.* Vols. 1-2. Chicago: University of Chicago.
Redfield, R.
1947 "The Folk Society". *The American Journal of Sociology* LII , no 4:293-308.
1960 *The Little Community and Peasant Society and Culture.* Chicago: University of Chicago.
Redford, D. B.
1992 *Egypt, Canaan, and Israel in Ancient Times.* Princeton, N.J.: Princeton University.
Reifenberg, A.
1955 *The Struggle Between the Desert and the Sown: Rise and Fall of Agriculture in the Levant.* Jerusalem: Government Press.
Robins, G.
1995 "Mathematics, Astronomy, and Calendars in Pharaonic Egypt" in *Civilizations of the Ancient Near East,* vols. 3 and 4, eds J. Sasson; J. Baines; G. Beckman; and K. Rubinson. Peabody, MA: Hendrickson. pp. 1799-1814.
Rowton, M.B.
1976 "Dimorphic Structure and Topology". *Oriens Antiques* 15:2-31.
Sansone, D.
2004 *Ancient Greek Civilization.* Oxford: Blackwell.
Sartre, M.
2005 *The Middle East under Rome.* Cambridge, MS: The Belknap Press of Harvard University.
Shahid, I.
1984a *Rome and the Arabs: A Prolegomenon to the Study of Byzantium and the Arabs.* Washington D.C.: Dumbarton Oaks.
1984b *Byzantium and the Arabs in the Fourth Century.* Washington D.C.: Dumbarton Oaks.
1989 *Byzantium and the Arabs in the Fifth Century.* Washington D.C.: Dumbarton Oaks.
1995 *Byzantium and the Arabs in the Sixth Century,* Vols 1 and 2. Washington D.C.: Dumbarton Oaks.
Sinopoli, C. M.
1994 "The Archaeology of Empires". *Annual Review of Anthropology* 23:159-180.
Smith, M.G.
1974 "The Comparative Study of Complex Societies" in *Corporations and Society: The Social Anthropology of Collective Action,* M.G. Smith. London: Aldine. pp. 241-365.
Steward, J.
1955 *Theory of Culture Change.* Urbana: University of Illinois.
Tristram, H.B.
1873 *The Land of Moab.* New York: Harper & Brothers.
Tritton, A.S.
1951 *Islam: Belief and Practices.* London: Hutchinson.
Tubb, J. N.
1998 *Canaanites.* Norman, OK: University of Oklahoma.
Veyne, P., ed.
1987 *A History of Private Life from Pagan Rome to Byzantium,* Cambridge, MS.: Harvard University.
Von Grunebaum, G.
1955 "The Problem: Unity in Diversity" in *Unity and Variety in Muslim Civilization,* ed. G. von Grunebaum. Chicago: University of Chicago. pp. 17-37.
Von Soden, W.
1994 *The Ancient Orient: An Introduction to the Study of the Ancient Near East.* Grand Rapids, MI: Eerdmans.
Wåhlin, L.
1997 "The Family Cistern: 3,000 Years of Household Water Collection in Jordan" in *Ethnic Encounter and Culture Change,* eds M. Sabour and K. S. Vikør. Nordic Research on the Middle East No. 3. pp. 233-249.
Walker, B. J.
1999 "Militarization to Nomadization: The Middle and Late Islamic Periods". *Near Eastern Archaeology* 62/4:202-32.
Wallerstein, I.
1976 *The Modern World-System: Capitalist Agriculture and the Origins of the European World-Economy in the Sixteenth Century.* New York: Academic Press.
Wetterstrom, W., and Murry, M. A.
2001 "Agriculture" in *The Oxford Encyclopedia of Ancient Egypt* vol. 1. Oxford: Oxford University. pp 37-44.
White, K.D.
1970 *Roman Farming.* Ithaca, New York: Cornell University.
Wittfogel, K.
1957 *Oriental despotism: A comparative study of total power.* New Haven: Yale University.

Wolf, E. R.
1997 *Europe and the People Without History.* Berkeley, CA: University of California.

Woolley, C. L.
1965 *The Sumerians*, New York: Norton.

Zurayk, C. K.
1949 "The Essence of Arab Civilization". *The Middle East Journal* 3/2: 125-139.

NOMADS AND CITIES: CHANGING CONCEPTIONS

Emanuel Marx

Introduction

In this paper I describe the gradual growth of my understanding of pastoral nomads in general and of the Bedouin of South Sinai in particular. Our conceptions determine which aspects of reality we see and which we ignore. Only a continuous dialectic between observations and analysis can overcome their innate resistance to change. It has taken me the best part of thirty years to evolve a conception of the Bedouin in South Sinai that comprehends the full range of phenomena I wittingly and unwittingly observed in the field. When I began the study I knew that Bedouin depend on cities. It took me some time to realize that they don't just depend on cities, but fully participate in urban civilization. It took me even longer to grasp that urban civilization is found in their midst, it pervades every aspect of their lives, such as territorial organization, the division of labour, patterns of mobility and consumption habits.

The First Conception

Between 1972 and 1982 I spent about twelve months in the field, mostly working with the Jabaliya Bedouin in the Mount Sinai region. It did not take me long to realize that they were quite different from the Negev Bedouin whom I had studied earlier (Marx 1967). With the Negev my theoretical starting point had been that before the advent of Israel the Bedouin had been sedentary and had participated in a diversified economy (Dan Gazit 2004). Then the Israeli state put the Bedouin under a Military Administration, confined them to a native reservation and did not permit them to participate in the wider economy. Thus they became almost ideal type Bedouin pastoralists and dry-land farmers. In Sinai such a conception did not work, for neither the Egyptian authorities, nor the Israeli occupying forces which replaced them, put much effort into administrating the Bedouin population. The Jabaliya did raise flocks and tended orchards but, in contrast to the Negev Bedouin, did not consider them a source of income. Theirs was a token pastoralism which, they often told me, "only causes losses, and that's why we try to keep our flocks as small as possible", and a horticulture that only produced fruit and vegetables for home consumption. They treated especially the fruit as "sweets" (*halawa*), and not as a substantial food. They worked mostly as labour migrants and claimed – and this is confirmed by documents stored in Saint Catharine's monastery (Shuqair 1906, Schwarz 1970, Humbsch 1976) and in travellers' accounts (Hobbs 1995: chapter 9, Mouton 2000: 145 ff.) that for many generations they had been labour migrants and caravaneers. They maintained the flocks and orchards as resources to fall back on in bad times. Experience had taught them that recurrent economic and political crises threw labour migrants out of work. Whenever there was a wave of unemployment the labour migrants were the first to lose their jobs. Then they would return home and reactivate their flocks and orchards, and thus rejoin the market economy by an alternate route. The flocks and orchards thus served mainly as an economic reserve, or as an insurance policy. During my fieldwork there occurred two crises which brought the reserve economy to life. One was the October 1973 war, which disrupted Israeli economic activities in Sinai for many months. The other, ironically, was precipitated by US Secretary of State Kissinger's shuttle diplomacy between Cairo and Jerusalem in 1976, which eventually resulted in a peace treaty between the two states and put an end to the Israeli occupation of Sinai. The Bedouin knew that these developments would usher in a period of unemployment.

Most Bedouin considered labour migration to be fundamentally unattractive, as it forced men to spend long periods away from family and home and also because it generally offered insecure menial work. While almost every Bedouin man spent many years of his working life as migrant labourer, he never intended to remain one for long. When his hopes to obtain permanent employment in the city were deceived, he could at least look ahead to the time when his sons would be old enough to take over, and he would be able to stay at home, in the mountains of South Sinai. Here he would cultivate his orchards and raise a small flock, and earn a small income that would be subsidized by the labour of his sons. In practice, these long-term projects hardly ever worked out as planned. For when work opportunities were abundant, the older men too would once again go to work as labour migrants, and when work was scarce – the flocks and orchards could not rapidly enough be converted into profitable enterprises.

For a long time I thought that labour migration was the key to understanding this society. It certainly was, and had been for generations, a major source of cash income. As labour migration typically offers only insecure employment, the Bedouin invested great efforts and resources in building up a reserve economy, in the widest sense of the word. The orchard and small flock, which each household maintained in a state of readiness for reactivation, was just one element of it. Others were the carefully fostered ties with kin and friends, the persons who could be relied upon to help in time of need; and the annual pilgrimage (see below), in which they renewed membership of the tribe, assured their right to plant orchards and build solid houses in the tribal area, to use the roads crossing the territory for commercial purposes, and to exploit pastures all over the region of the tribal confederation. While the men were away at work, the women and children operated the reserve economy at low intensity, barely keeping the orchards and flocks alive, and also kept up relations with kin and friends. The men too participated in this maintenance work: they would return home for short vacations, during which they exerted themselves by visiting kin and friends and bringing them small gifts. When the men returned home for extended terms, they turned the orchards and flocks, by dint of hard work, into profitable ventures. Toward the date harvest in autumn many men returned home, usually for several weeks, in order to take their families on the annual tribal pilgrimage to the tombs of the tribe's ancestor and other saints. That was the only time when the members of the tribe, who all year had been scattered among numerous distant workplaces, were reunited. Together they re-embodied the tribe and reaffirmed its territorial rights, while each individual consolidated his or her relations with the saints, the intercessors with a provident god.

The Second Conception

This was the conception I used in my first articles on the Mount Sinai Bedouin (Marx 1977, 1980, 1984, 1987), and it seemed to explain much of the data. But soon I was writing papers that no longer fitted this conception. The change was precipitated by a comparative study of pastoralists in the Arab Middle East that I began at the time (Marx 1992, 1995). A major finding of that study was that pastoralists are so deeply involved in the market economy that they should be viewed as specialized segments of a complex city-based economy. In the city they market their produce, such as meat, hides and wool, and buy all the necessities of life, including most of the grains that are their staple food. It is also the major centre for work and services, and the seat of government. They depend so heavily on cities, that any economic or political turbulence there may change their entire way of life, even to the point that they temporarily or permanently cease to be pastoralists. Such traumatic events, I realized, recurred quite regularly in the course of the peninsula's history. Being the land bridge between Africa and Asia, it was incessantly invaded by migrants and overrun by armies from North and South.

The new conception led me to re-examine my earlier work. I realized that the Bedouin were even more profoundly involved in world events than I had thought, and that my writings had not portrayed all the complexity and adaptability of their society. In particular, the economy of the South Sinai Bedouin was more diversified than I had allowed for. The re-examination of the ethnographic material resulted in three articles. The first paper (1999) examined the various types of oases found in Sinai. It touched tangentially on the importance of smuggling of narcotic drugs to the Bedouin economy. Before the Israeli occupation, smuggling had driven economic development in South Sinai, and smuggling bands had even created several new oases, as hideouts in inaccessible mountain areas. While smuggling was at low ebb during my fieldwork, it remained an important, though submerged, aspect of Bedouin society.

A second paper (2004) dealt with roving traders and their critical contribution to the economy of the South Sinai Bedouin. The traders are the long arm of the city: they supply the Bedouin with grain, and almost all the other necessities of life. While grain is the Bedouin's staple food, it is not grown in the region, and had it not been for these traders the Bedouin simply would not have survived in arid Sinai. Under Egyptian rule, this trade had been in the hands of Bedouin cameleers and truckers, who worked closely with Egyptian storekeepers in Suez. Only during the Israeli occupation was the trade taken over by merchants from the town of al-'Arish in North Sinai. Finally, I wrote a paper (in press) which no longer viewed labour migration as the major or only factor shaping society. The reserve economy, I realized, was not simply a response to the exigencies of labour migration, but was a way of life in its own right. Philip Salzman (1980: 7-10) had tried to convert me to this viewpoint, but at the time I did not heed his good advice.

I was able to write the present paper, the third in the series, only when I realized that the Bedouin not only participated in the great society and its market economy, but that it was an integral part of their local society. This was evident in a diversified occupational structure which, of course, I had seen. It was hard to miss. But I had not discussed it in my work, because I still viewed the Bedouin as pastoralists and labour migrants. True, practically all Bedouin households raised flocks and tended orchards, and most men were labour migrants, and quite a few were former and potential smugglers. But many men also engaged in a specialized trade. They were well-diggers, builders, fishermen, caravaneers, truck drivers, car mechanics, and vegetable growers. Several men produced the local variety of tobacco and thus infringed a state monopoly. They were pollinators of date palms, fishermen, boat builders, shopkeepers (*kantin*), and owners of coffee-houses. They were government employees, such as tribal chiefs, teachers and informers. There were also tribal judges, healers, jewellers, tailors, basket makers, tourist guides, and permanent employees (*subyan al-deir,* the lads of the monastery, a term which tribesmen also applied to all the Jabaliya) of Saint Catherine's monastery. Nearly all these specialists were men, but there were women healers, fortune tellers and midwives. All transactions with these specialists were paid in cash.

This specialization was even reflected in the competences of the local healers and in the special powers attributed to each saint. Each healer treated certain complaints; one was an expert on diabetes, another specialized in circumcisions and extractions of the uvula, operations to which every boy and girl was subjected, and another again treated mental problems. Saints too were thought to deal efficiently with some problems and not with others. Bedouin often approached one saint when their children fell ill, barren women visited another, a third saint could bring back a lost camel, and yet another assure the welfare of the flocks.

This does not imply that the economies of the city and of the Bedouin were almost identical. The low population density, the very limited literacy and technological training of most of the inhabitants, the peculiar combination of labour migration, pastoralism and orchard cultivation, all these produced a special brand of complex economy. It required the inhabitants to cover great distances in order to conduct routine economic transactions; it induced them to use locally available materials, and to exercise great ingenuity in operating modern technologies. Thus a visit to a healer, often combined with a personal pilgrimage to a saint's tomb, could take several days of travel by foot or by car. The basket-maker al-Rifi of Wadi Firan used only local raw materials, the fronds and fibres of palm trees. Even when industrially produced cords were available, he preferred to twist his own ropes. The items he produced were also different from those found in the cities and villages of Egypt: he made containers for dates, baskets for coffee-making utensils or for clothing (to be hung from the rafters of the house or roof of the tent), chicken-coops and a kind of incubator for new-born chicks. As the local car-owners had no access to spare-parts, they carried out ingenious improvised repairs with odd pieces of wood and ends of barbed wire left behind by the armies passing through Sinai.

Furthermore, "Bedouin" society included many strangers and outsiders. The roving merchants from al-'Arish obviously were part of it, as they made it possible for the Bedouin to live in the region. But so were the tourists, who brought cash to the people who stayed at home, the government officials who provided essential services, the soldiers who interfered in the Bedouin's way of life and drove up the price of narcotics, and the Greek monks who inspired the Bedouin's horticulture, stored their important documents in the monastery, and supported their claims to land ownership before the authorities.

Thus I came to treat Bedouin society as a complex and economically differentiated social agglomerate, whose members participated in various ways in the market economy. I also dealt with the impact of rapid and often turbulent changes in the wider political arena on the Bedouin economy. The study in its present state shows how various economic pursuits produced numerous sectors and lifestyles within this Bedouin society, existing side-by-side and yet linked up with one another. It became quite unlike the standard anthropological monograph. It used an open-system analysis and thus emphasized wider social contexts. The ethnography became exceedingly complex and disjointed, and thus lacked the dramatic unity of persons, time and plot that makes a good read. The account of the tribal pilgrimage (2005) that concludes the study, however, reflects and fuses many of these strands of society.

The Bedouin of South Sinai illustrate one way of making the profound connection between nomads and cities: they are an attenuated version of the city's complex economic specialization and differentiation. Even so, they still make numerous excursions to the city, in order to benefit from the full range of its services. The Syrian Rwala Bedouin today (Lancaster 1997: 166-7) exemplify a very different type of reliance on cities: their major specializations in sheep-raising, commerce and trucking require constant moves between cities and markets in four countries (Saudi Arabia, Iraq, Jordan and Syria). But they too have for generations had numerous specialized artisans and traders living in their midst, while the tribe's chiefs reside almost permanently in the capital Damascus.

In conclusion, I suggest that the conception presented here applies to many social aggregates in which pastoral nomadism plays a significant economic role. The city's diversified economy lives in their midst.

Acknowledgements

I read drafts of the article at a workshop on Pastoral Markets in East Africa and the Middle East in Pisa, July 2003, and a conference on nomads and settlers in Wittenberg, November 2003. I thank Gudrun Dahl, Klaus Ferdinand, John Galaty, Anatoly Khazanov, Joram Kufert, Peter Little, Michael Horowitz, Frank Stewart, Eveline van der Steen and Richard Werbner for many useful comments.

Bibliography

Hobbs, J. J.
1995 *Mount Sinai*. Austin: University of Texas.
Humbsch, R.
1976 *Beiträge zur Geschichte des osmanischen Ägyptens: nach arabischen Sultans – und Statthalterurkunden des Sinai-Klosters*. Freiburg: Schwarz.
Lancaster, W.
1997 *The Rwala Bedouin Today*, 2nd edition. Prospect Heights, Illinois: Waveland.
Marx, E.
1967 *Bedouin of the Negev*. Manchester: Manchester University.
1977 "Communal and individual pilgrimage: the region of saints' tombs in South Sinai" in *Regional Cults*, ed. R. P. Werbner. ASA Monograph, no.16. London: Academic Press. pp. 29-51.

1980 "Wage labor and tribal economy of the Bedouin in South Sinai" in *When Nomads Settle: Processes of Sedentarization as Adaptation and Response,* ed. Ph. C. Salzman. New York: Bergin. pp. 111-23.

1984 "Changing employment patterns of Bedouin in South Sinai" in *The Changing Bedouin*, eds E. Marx and A. Shmueli. New Brunswick, New Jersey: Transaction. pp. 173-86.

1987 "Labour migrants with a secure base: Bedouin of South Sinai" in *Migrants, Workers, and the Social Order*, ed. J. S. Eades. ASA Monograph, no. 26. London: Tavistock. pp. 148-64.

1992 "Are there pastoral nomads in the Middle East?" in *Pastoralism in the Levant*, eds O. Bar-Yosef and A. Khazanov. Madison: Prehistory Press. pp. 255-60.

1995 "Are there pastoral nomads in the Arab Middle East?" in *The Anthropology of Tribal and Peasant Pastoral Societies*, eds U. Fabietti and Ph. C. Salzman. Pavia: Collegio Ghislieri and Como: Ibis. pp. 71-85.

1999 "Oases in South Sinai". *Human Ecology* 27(2): 341-57.

2004 "Roving traders among the Bedouin of South Sinai" in "*Customary Strangers*", eds J. C. Berland and A. Rao. Westport, Connecticut: Praeger. pp. 57-69.

2005 "Tribal pilgrimages to saints' tombs in South Sinai" in *Archaeology, Anthropology and Cult: The Sanctuary at Gilat (Israel)*, ed. Th.E. Levy. London: Equinox.

In press "Nomad and City" in *Shifts and Drifts in Nomadic-Sedentary Relations*, ed. S. Leder.

Mouton, J.-M.

2000 *Le Sinaï medieval: un espace stratégique de l'Islam*. Paris: Presses Universitaires de France.

Schwarz, K.

1970 *Osmanische Sultansurkunden des Sinai-Klosters in türkischer Sprache*. Freiburg: Schwarz.

Salzman, Ph. C.

1980 "Introduction: processes of sedentarization as adaptation and response" in *When Nomads Settle: Processes of Sedentarization as Adaptation and Response*, ed. Ph. C. Salzman. New York: Bergin. pp. 1-19.

Shuqair, N.

1906 *Ta'arikh sina al-qadim wal-hadith wa-jighrfiataha* (The History and Geography of Ancient and Modern Sinai). Cairo: Ma'arif.

SETTLEMENT PATTERNS IN THE LATE BYZANTINE AND EARLY ISLAMIC PERIODS IN THE NEGEV, ISRAEL

Dov Nahlieli

Introduction[1]

In recent years archaeological research carried out in Southern Israel in the Western Negev, Beersheva basin, Northern Negev highlands, the Central Negev highlands and the Arava and Eilot-Eilat region has revealed valuable information concerning settlement patterns in the Late Byzantine and Early Islamic periods (5th-9th century CE). Much of this research has been facilitated by archaeological surveys of large areas of the Negev, particularly in the Central Negev in the region of Shivta and Sede Boqer, as well as further south in the area of Har Hamran and the area southwest of Mizpe Ramon (Cohen 1981, 1985, Baumgarten in press, a, in press, b, 2004. Haiman 1986, 1991, 1993, 1999; Rosen 1985, 1994). In addition, data has been obtained from excavations and surveys of regions situated further north in Nahal La'ana, Ramat Divshon, and Har Nafha as well parts of the Ramon Crater and the region around Har Sagy (Nahlieli 1999; Applebaum 1966, Lender 1990a, Rosen 1994, Avni 1985, 1992).

These surveys and excavations reveal that the number of sites dating to the Byzantine period is larger than that of any other period in the Negev.[2] These findings match similar evidence found in other parts of Israel and Jordan. An overall examination of the evidence shows that the population and the number of settlements throughout this part of the Middle East peaked during the Byzantine period between the 5th and 7th century CE (Tsafrir 1996: 269-283).

Settlement Types in the Negev Highlands

Systematic map surveys carried out in the Northern Negev highlands have identified the following five types of site described below:

1) Large central villages (for example, Shivta, Rehovot-in-the Negev, Nessana).
2) Farmsteads containing square buildings constructed from roughly dressed stones. These structures contain a number of rooms and one or more courtyards. Farms of this type are usually located in agricultural areas next to terraced wadis.
3) Farms containing a single, square structure constructed from dressed stones. These structures share distinctive architectural features, i.e. entrances constructed with dressed lintels built next to a corner of the structure, dressed thresholds and occasionally arches designed to support stone ceilings. Structures of this type usually contain nearly identical dimensions of 4 x 4 m. although a few have been found with somewhat larger or smaller dimensions. These structures are generally found located next to terraced wadis near large, central settlements.
4) Temporary sites containing structures constructed from rough field stones. Structures found at these sites often contain a number of rooms and one or more courtyards. This type of site often consists of a number of occupant structures as well as animal pens. Like the farmsteads, the majority of these sites are found located next to terraced wadis and a number have been found next to cisterns and threshing floors.
5) Sites which do not contain signs of permanent structures but are situated next terraced wadis, cisterns and threshing floors.

Settlement patterns in the Byzantine and Early Islamic Periods in Other Parts of the Negev and Arava Valley

The settlement picture in the Beersheva basin is essentially similar to that of the Central Negev highlands. Surveys and excavations have uncovered farmsteads from the Byzantine period which existed in close proximity to the town of Beersheva. The inhabitants of these farms belonged to communities of farmers who lived on the periphery of the town and supplied it with agricultural products. Due to their proximity to Beersheva the inhabitants of the farmsteads appeared to have enjoyed a measure of prosperity, as can be detected in the relatively rich assemblages of finds not commonly found in farmsteads in the Negev highlands (P. Fabian, p.c.). Like those in the Negev highlands, the farmsteads in the Beersheva basin continued to be inhabited in the Early

[1] I would like to thank Dr. Tali Erickson-Gini for assistance in translating this material from Hebrew into English and also for her help in preparing this paper for publication

[2] Researchers relying on material obtained from archaeological surveys have encountered some difficulties in distinguishing the exact dates of farm settlements in the Negev probably due to the existence of a transitional period between the Late Byzantine and Early Islamic periods. Tsafrir has demonstrated the problem of dating these farmsteads: in summarizing data obtained from surveys made up until the publication of his article he concluded that they were constructed in the Byzantine period and continued to be occupied in the Early Islamic period (Tsafrir 1996:272). I agree with Tsafrir's findings that farmsteads in the Negev Highlands were constructed at the end of the period of Byzantine rule.

Islamic period (Ustinova and Nahshoni 1994: 170). Another site located north of Beersheva, at Khirbet 'Amra also belongs to the Byzantine period. Excavations at this site have uncovered at least one church, while aerial surveys have revealed evidence of a second church. However, in the agricultural areas near the settlement farm structures were found belonging to the Byzantine period and primarily to the Early Islamic period (G. Tahal p.c.). In recent excavations north of Beersheva Fabian and Feder found a farmstead site containing several rooms and single roomed structures (Fabian and Feder, p.c.). These investigations have revealed a feature similar to that found in the Negev highlands: the extensive farmsteads that have been dated to the Early Islamic period appear to have incorporated the Byzantine watch tower or "shomerot" in their construction (Nahlieli 1999: 77*).

Research carried out in the Western Arava region has revealed a completely different picture from that found in the Northern and Central Negev. With the exception of Roman military installations such as the one at Mezad Ein Hazeva, this area contains virtually no sites dating to the Byzantine period. However, several sites belonging to the Early Islamic period have been surveyed and excavated in this region (Israel *et al.* 1995, Nahlieli *et al.*, f.c.; Avner 1998). These sites contain very simple structures and generally consist of one or two rooms with a courtyard. Additional rooms are usually arranged in rows. The foundations are constructed using smooth wadi stones and the walls are built of mudbricks (Israel *et al.* 1995: 1; Avner 1995: 105). Agricultural sites in the Arava belonging to this period include "kanat" systems at Ein Yahav, Yotvata and Ein Avrona, and aqueducts, features not found in the Negev Highlands (Porat 1987: 106).

The Nature of Early Islamic Settlement in the Negev: Farmers or Herders?

In recent years, two main theories have been proposed to explain the existence of Early Islamic settlement in the Negev. The first, proposed by Haiman, suggests that Early Islamic settlement activity was a policy of the Umayyad regime in which seminomadic groups in the region were encouraged to settle frontier areas that were vulnerable to seminomadic raids and invasions. Haiman attributes the extensive construction of terraced farms found throughout the Negev to this period in spite of the current consensus among researchers that they date to the Byzantine period (Haiman 1995: 39-41). The second view, proposed by Avni, is that of sedentarization of seminomadic elements living on the fringe of the Byzantine towns in the Early Islamic period when these towns were in decline (Avni 1996: 86-98). According to this theory, the sedentarization of seminomadic groups in the area in this period was a natural process and a response to changing economic realities in which these groups adopted a sedentary lifestyle in order to compensate for the loss of their original means of livelihood, which was heavily dependent on the prosperity of the urban population of the Negev.

In contrast to the views expressed by Avni and Haiman concerning the nature of Early Islamic settlement in the Negev Highlands described above, I propose that these sites did not belong to the nomadic inhabitants in the area.[3] Nomadic populations dependent on sheep and goat herding would not have had the knowledge or ability needed to maintain the extensive system of terraced farms located throughout the Central Negev highlands. It should also be pointed out that the large central settlements of the highlands would have required extensive agricultural lands in order to make a livelihood in such an arid region.[4]

Throughout the Middle East in this period the status of urban centres declined as the result of a loss of economic control and administrative changes which began in the second half of the sixth century CE. Plague, earthquake damage and other disturbances (Foss 1997: 189, Conrad 1986: 146) added to this process of decline. These events affected the region of southern Israel and the Negev towns as well. Numismatic and other material evidence obtained in excavations in the Ashkelon region at the site of the "Third Mile Estate" and the nearby site of Hamama points to the abandonment of those sites in the second half of the sixth century CE (Ariel in press). The abandonment of the prosperous "Third Mile Estate" is enigmatic and can possibly be attributed to the occurrence of plague at the site (Israel and Erickson-Gini, in press).[5] Studies carried out on the distribution of epitaphs in the western and central Negev have revealed evidence that the plague first appeared in southern Israel in coastal ports in the early spring of 541 and progressed into the Central Negev towns by autumn (Di Segni 1997: 911). Epitaphs discovered in Southern Israel and in Moab have confirmed evidence of a second wave of the plague in the region of the Eastern Mediterranean a decade later in 554/555 CE as reported by Evagrius Scholasticus (Di Segni 1997: 911-912, n. 3). In addition to the plague, documentary evidence dated to the sixth century points to difficult climatic conditions, probably drought and sandstorms which appear to have had a detrimental effect on the agricultural regime in the area of Elusa (modern Halutza), the largest town and provincial capital in the Byzantine period. One contemporary source, Procopius of Gaza (c. 450-526) described the encroachment of sand dunes around the city, as well as drought (Mayerson 1983: 251-252). Mayerson points out that further evidence of the encroachment of sand dunes into the

[3] Regarding Magness' recent publication on Early Islamic settlement in Palestine (2003), I disagree with her assertion that no decline took place towards the end of the Byzantine era in the Negev. However, it is beyond the scope of this article to treat the subject fully (D.N).

[4] Even today it is possible to find inhabitants from towns and villages in the region of Judea and Samaria who utilize some areas of the Judean Desert for agriculture. During the agricultural season these families leave their permanent homes and take up residence in temporary sites such as ruins, caves and campsites in order to sow and harvest crops

[5] Excavations in the "Third Mile Estate" revealed evidence of a prosperous agricultural estate complete with two marble lined winepresses, an industrial sized kiln that produced "Gaza" wine jars for export, extensive storerooms, facilities for producing olive oil, a luxurious bathhouse and possibly fish ponds.

coastal area has been found at other sites in the region. At Deir el-Balah west of Gaza 13 meter high sand dunes were found overlying Byzantine sites (Mayerson 1983: 253).

Destruction wrought by earthquakes has been detected at several sites in the central Negev. Avdat (Oboda) was apparently totally destroyed by an earthquake and subsequently abandoned in the early seventh century (Fabian 1996). The nearby Byzantine village of Mezad Yeruham also appears to have been destroyed and abandoned in the same event (Y. Baumgarten, p.c.). This earthquake appears to have affected most of the large towns in the central Negev, as can be seen by the construction of revetment walls around churches and private dwellings at Shivta, Rehovot in the Negev, Sa'adon and Nessana (Erickson-Gini f.c.).[6] It is probable that, faced with deteriorating conditions in the sixth century, inhabitants who could abandon the area did so, but for the majority of the local population this option was not available, and many inhabitants moved out of the central settlements and into agricultural areas on the periphery which was formerly inhabited on a temporary, seasonal basis. There they built their homes. In the initial phase structures were built containing only one or two rooms. Later, and according to need, several rooms were added to the original dwelling.

The settlement pattern in the central Negev began to shift from towns, which the inhabitants left temporarily in order to work the fields in the vicinity of the towns to widely dispersed groups of people spread out over many kilometers. As a result, a few of the larger towns and villages, such as Avdat (Oboda), Mezad Yeruham and Mamshit (Mampsis), went out of existence and were replaced by several dozens of farmsteads occupied by extended families. This change in settlement pattern is particularly noticeable in the region around Avdat and Sede Boqer where several Early Islamic sites have been discovered. These sites were most likely inhabited after the town of Avdat was destroyed by an earthquake in the early seventh century CE. Three towns, Shivta, Nessana and Rehovot in the Negev, continued to be inhabited into the Early Islamic period. At Shivta, the elaborate system of dams and terraced walls in the Nahal Lavan and the less elaborate terraced fields in Nahal Zeitan and other nearby wadis appear to have been repaired and used in the Early Islamic period just as the ancestors of the local inhabitants had done over a century earlier.

Historical Considerations: The Muslim Conquest of the Negev

Until the time of Justinian settlement in the Negev the Byzantine period was characterized by a concentration of the population in towns, which were relatively large in comparison to settlements in the Negev in other periods. This picture changed, in my opinion, towards the end of the sixth century CE and prior to the Moslem conquest. At that time the large settlements suffered a decline in central administrative influence and the inhabitants suffered a major loss of income when local soldiers, the limitanei, were deprived of their traditional payment (Negev 1993:241). In spite of this fact, it appears that the authority of the clergy, or more likely that of the local bishop, remained intact (Haldon 1990: 97). This situation came about as a result of disturbances which took place during the reign of Phocus, and particularly as a result of the Persian invasion which cut off South-West Asia from Asia Minor, and during which period, after 603 CE, the area was not under active Byzantine control. During the Moslem conquest of the area, the Islamic army negotiated terms of surrender with Church officials in local settlements throughout the region. This was the situation in Ayla, which was surrendered by Yuhanna b. Ru'ba, the local bishop (Ibn Sa'd A (2): 28-29; Tabari Ta'rikh, A: 1702, Waqidi: 1031; Baladhuri: 60). The security of the southeastern periphery of the Empire was placed in the hands of the Ghasnids and local tribesmen, particularly in the region of the Golan and Transjordan. Tribesmen in these areas were willing to guarantee the security of the region in return for special privileges and monetary grants (Gil 1983: 15). The settlements in the Negev already enjoyed close association with these tribes and it is reasonable to assume that at least a part of the local population were made up of tribe members. Cooperation between local settlements and tribesmen included defense agreements and treaties (Donner 1989: 82).

This situation is comparable to that in Syria, where the Moslems took over Syrian towns following negotiations with little or no opposition to the conquest appearing to have been exhibited by the inhabitants. The population was required to pay levies to the Moslem authorities and negotiations were carried out along the lines of similar agreements commonly made between the settled population and nomadic tribesmen prior to the conquest. The nomads agreed to protect the settlements and allowed them to carry on their way of life in return for various payments. Syrian Byzantine sources do not appear to have viewed the invaders as a new religion or unknown element. According to these sources, the Moslem invasion was viewed as a temporary incursion brought about to punish the Christians for their sins (Brock 1982: 14; Palmer 1993: xiii). It appears that agreements of this type between the local settled population were well known in the periphery of the Empire before the introduction of the Moslem armies and were accepted practice between the sedentary population and Beduin tribesmen. These agreements created the unusual situation in which the inhabitants of the south-east periphery of the Empire enjoyed a relative measure of stability at a time when the Empire itself was suffering from major instability.

Both in Syria and in the Negev the local inhabitants now enjoyed the benefits of the security provided by local

[6] Although revetment walls have not been found inside the Byzantine town of Mampsis, one such wall has been discovered along the northern exterior of a chapel located in the Byzantine cemetery at the site (T. Erickson-Gini, p.c.). Other possible signs of earthquake damage may include the apparent abandonment of many private dwellings at Shivta, the doorways of which have been blocked (Erickson-Gini, f.c.).

nomadic tribesmen. It must be kept in mind that the same population had recently suffered the effects of plague, earthquake, taxation and the loss of military payments and they were eager to improve their situation. Their major source of income was inevitably based on the cultivation of their land, and they were unable to abandon their property and escape. The only solution for many inhabitants in the larger towns was to transfer their families to the cultivated areas and set up their permanent homes there. Their proximity to the fields allowed them to work the land intensively. In the sixth and beginning seventh century CE, the city of Antioch suffered badly from earthquake, plague and the Sassanian invasion and yet the village area surrounding the city continued to flourish (Foss 1997: 190; Kennedy and Liebeschuetz 1988: 87). This makes clear that this process of a withdrawal from urban centres to village areas is not restricted on to the Negev.

The Population of the Negev in the Early Islamic Period

The settlements of the Negev in the Early Islamic period were generally small villages or farmsteads that appear to have been initially established earlier in the Byzantine period. At first, the shift from Byzantine rule to an Islamic administration did not bring about apparent changes in the makeup of the local population. In the wake of the consolidation of the Arab system of taxation and administration, many local landowners were forced to give up their private property and were turned into virtual serfs on their land.

During this period the inhabitants of the Negev were employed in the same occupations that were common earlier in the Byzantine era. These occupations were common in antiquity and the Negev was not different in this respect from other regions. The inhabitants' major sources of income were agriculture and animal husbandry. The settlers in this region overcame serious obstacles such as a lack of water sources and cultivable land. Their amazing success in producing crops in this arid zone enabled them to survive and prosper in an environment normally unsuitable for permanent habitation.

Archaeological evidence for the latest occupation of many Early Islamic sites points to the abandonment of the area by permanent settlers, who were replaced by nomadic elements moving into what were formerly cultivated areas.[7] Initially it appears that the nomadic population continued to work the formerly cultivated land. However, the area eventually became deserted when the permanent settlements ceased to exist and only Beduin tribes continued to inhabit the area.

In the Arava Valley the Arabs created three new agricultural centers at Ein Yahav, Yotvata and Ein Avrona, which each became a virtual oasis. These sites were probably constructed under the official sponsorship of the Umayyad regime as part of a three pronged policy that included the development and expansion of copper mining at Timna and Feinan, the development of agricultural products for consumption by the population of Ayla and the Muslim pilgrims on the pilgrimage routes, and the construction of stations along the main route leading north from Ayla to Ramle.

The Administration of the Negev in the Early Islamic Period

From an administrative point of view the Arabs conquered a well-organized land. The Byzantine administrative system was staffed with professional administrators (including officials of the clergy). It is not clear how the new administration was organized immediately after the conquest, however, from the reign of Mu'awiya (the founder of the Umayyad dynasty) it is evident that the administration adapted the old bureaucratic institutions. The Arabs adjusted the former Byzantine administration to suit their own needs. The Caliph, Abd el-Malik first began a process of modification of the administration. He "decided that the time had come for Arab rulers to understand the administrative actions of its officials; and accordingly he ordered that all important documents be translated into Arabic, including matters of finance and taxation, and from this time all official records were written in Arabic." (Cahen 1995: 559). This was a gradual process: Christian officials were replaced by Moslems (some of which had converted from Christianity to Islam) and the Arabic language became the official language of the Caliphate.

In the Early Islamic period three main phases of consolidation in the Arabic system of taxation have been documented. In the first phase, the Muslim authorities continued to employ the Byzantine system of taxation that was formerly used by Church officials. Levies were imposed on communities as a whole. In the second phase a change in the administrative system of taxation was instituted and a new system was established in which individuals were taxed according to the size of their property holdings and ability to produce. In the third phase taxes on property holdings were transferred directly to the Arab tribesmen. This latest form of taxation is not indicated at all in the documents dated to the later seventh century in the Nessana archive, and it probably dates to the period after the mid-eighth century. Prior to that the mid-eighth century revenues were evidently divided up by the administration in Gaza after collection.

The Abandonment of the Negev under the Abbasid Administration

In the wake of the fall of the Early Islamic government and the accession of the Abbasid regime the existence of the Arava settlements was short-lived. This change in rule brought about a shift of the centre of power to Iraq, resulting in the loss of sponsorship of agricultural settlement in the Arava area, a region in which existence is quite difficult to maintain. The inhabitants were now

[7] Substantial layers of organic material have been found by the writer, overlying the latest phase of occupation in Early Islamic sites in excavations.

forced to abandon their homes and fields and the area became uninhabited.

Similarly, the agricultural settlements in the Negev highlands and Beersheva basin were a short-lived phenomenon. It is apparent that by the end of the seventh century CE the economic demands that were placed on the inhabitants living in those areas, in the form of heavy taxation, created great difficulties. In order to obtain relief the heads of several key settlements organized a delegation and brought their pleas before the Moslem governor. This delegation was formed by means of a letter carried from one settlement to the next (Kraemer 1958: 213, Document No.75). Although the actual effect of this letter on the administration is unknown, archaeological evidence shows that the settlements continued to exist for several more years (e.g. Haiman 1990: 122; Lester 1990: 125). Of course it is reasonable to assume that some families gradually abandoned the area due to economic hardship. Baladuri quotes a source which claims that in the days of Haarun Al-Rashid many farmers abandoned their land because of the high tax on land, the "Haraj" payments (Baladuri Futuh: 171). Another source found in Tabari (Ta'rikh: 711) mentions a revolt led by a man named Abu Al-Nida in the year 191 (Hijra). Ibn al-Athir (al-Kamil: 205, 208) refers to this man and claims that Yahya ben Maad brought Abu Al-Nada to Al-Rashid for execution. Ibn al-Taghribirdi, evidently basing himself on Tabari, further enlightens us on this matter and explains the reason for the revolt (al-Najum II: 135, 141): "When al-Hussein set the Haraj payment it became difficult to bear, and the people of Egypt came out against him and refused to pay the haraj..." Gil views this event as an actual revolt (Gil 1983: 241-242), however, the sources appear to refer to Abu el-Nida as a common highwayman and robber. In any case, the fact that the revolt was related to the matter of taxation is in itself evidence of the hardship that taxes caused the local population in an area in which survival was already difficult. It appears that in the end taxation and nomadic Saracen incursions were finally the "straw that broke the camel's back" and ended the period of agricultural settlement in the Negev. In the Chronology of Theophanes (died approx. in 818 CE) he wrote that in the years 812-813, " In the same year that many monks and the Christian priests left the Land of Israel and Syria for Cyprus, fleeing from the Arabs who know no border, and of the general chaos that rules in Syria, in Egypt, in Africa and over the entire empire of the Arabs..." (Griffith 1993:62). Other sources tell of the emigration of monks and local inhabitants, adding to the evidence we have concerning the economic pressure brought on the population by government authorities during a period in which the small settlements of the Negev were abandoned.

The abandonment of these settlements in the Negev during this period was caused by a sequence of events known to us from earlier periods of settlement in the region, which has been demonstrated by Cohen to have taken place in the Chalcolithic through the Iron ages in the Negev highlands (Cohen 1999, Cohen and Cohen-Amin 2004). This sequence of events is that of a cessation of external support, usually of a central government which originally subsidized activity in the region and established security and stability necessary for successful habitation. Inhabitants in this arid region have constantly been forced to combat shortages of water and difficult climatic conditions in an area located far from economic centres, as well as saline soil and a lack of fertile farmland. In addition, the settled inhabitants had to deal with the harassment of nomadic tribes and with unsympathetic administrations committed to raising debilitating taxes, wrecking the basis for the survival of settlements and forcing their eventual abandonment.

During the Roman and Byzantine periods the region served as an active land bridge for the Arabian peninsula. The area flourished and settlements were established along its major routes. In the wake of the establishment of the Islamic Empire and its control over far-flung areas such as Spain, North Africa, the Middle East and a large part of Asia, major routes of travel ran north towards the large population centres of the East. Under Umayyad rule, the government collected revenues from the populations of Syria and Israel, the rulers had a vested interest in the agricultural activity of the region and consequently they imitated the Byzantine administration in defending the settled population and blocking the desert (Sharon 1976: 21). However, in the wake of the Abbasid takeover the centres of administration were transferred eastwards and official policy changed, now favouring the eastern provinces over those of the west. This new situation brought about the neglect of the immediate region. "In the beginning this was expressed by the lack of interest shown by the government concerning irrigation systems and local defence of the sown regions. As the cultivated regions became abandoned the nomads increased" (ibid: 25). This same process has taken place in the desert in earlier periods, according to Biblical accounts (Judges 6: 2-5).

During the stages of abandonment religious nomadic elements entered the Negev and left behind inscriptions found throughout the area. In my opinion these inscriptions were not created by farmers living in the region. No inscriptions of this type were found adorning any residential structures and in addition no inscriptions thanking Allah or asking for his protection over the owner of a household were found. These inscriptions are invariably found on rock outcrops, on the remains of stone fences or on the remains of public buildings such as churches in Mampsis and Rehovot-in-the-Negev (for Mampsis see Sharon, in press; for Rehovot-in-the-Negev see Lecker 1989: 24; Tsafrir *et al.* 1988: 187; Sharon 1993: 50) and even on agricultural terraces (Sharon 1981: 87; Sharon 1985: 90; 1990: 9 and many other inscriptions presently being published by Sharon). The inscriptions found scratched on the churches were undoubtedly created after the churches' abandonment and destruction.

The inscriptions are, "for the most part stereotypical, repeatedly using the formula, 'Oh Allah forgive ...'" (Sharon 1990: 81, 85, 90). However, it should be stressed that they were inscribed by talented calligraphers and a few show knowledge of passages of the Koran (Sharon 1990: 16). Although it is as yet unknown who created these inscriptions or why they are found in this particular area they provide two fundamental pieces of information: the presence of Islamic elements, together with specific dates in some cases, and the atmosphere of the period in which a great deal of enthusiasm was expressed for the new religion of Islam, which included Messianic expectations.

In several Early Islamic farms a later occupational layer was found overlying an earlier one. This later layer was usually made up of a great deal of organic material including large quantities of goat dung (Nahlieli *et al.* 1996: 70). The archaeological evidence in this late occupational layer was very sparse and could not be dated. However, carbon-14 examinations indicate that during the 10th century CE a new population made use of the abandoned structures. This population was nomadic and used the abandoned structures as sheepfolds and pens and possibly also for human habitation. It may have been part of a second wave of nomadic emigration from the Arabian Desert northwards. This wave was made up of tribes such as the Benu-Halal and the Banu-Salim who penetrated Sinai by way of the Arava and from there reached Egypt (Sharon 1977: 200). It is possible that some tribes lingered in the Negev and herded their flocks there for a brief time. This population took advantage of the cultivated areas in which agricultural terraces and cisterns were found and may even have partially utilized them, not unlike modern day Bedouin. However, this was not the intensive, sophisticated agriculture found in the Negev in the Byzantine period and earlier part of the Early Islamic period. From the archaeological evidence it would appear that these nomads did not stay long in the area and after a few years they continued northwards or westwards.

Conclusion

The evidence presented here points to a process of deterioration of the large settlements of the Negev beginning as early as the second half of the sixth century CE. The inhabitants of these settlements abandoned a number of towns but did not abandon the area. They relocated their families in the cultivated areas throughout the Negev highlands and built farmsteads. This process was not directly encouraged by the Moslem authorities and may even have begun prior to the Moslem invasion as a result of a series of disturbances and crises which generally afflicted urban centres in the Middle East in this period. Following the Moslem invasion the local population in the Negev continued to survive in the area as long as the central government provided for their security. However, following the rise of the Abbasid regime and the transfer of power and interest to the East, the Negev was soon abandoned by its permanent inhabitants and the desert reclaimed the region that had been transformed into a flourishing garden by its former industrious inhabitants.

Bibliography

Ancient Sources

Al-Baladhuri, Ahmad ibn Yahya.
1956 *Futuh al-Buldan*. Cairo.

Ibn al-Athir, Ali b. Muhammad.
1965 *Al-Kamil fi alta'rich*. Beirut.

Ibn Sa'd.
1905-1940 *Kitab al-tabaqat al-kabir*. Leiden.

Ibn Taghribirdi, Abu al-Mahasin Yusuf
1929-1956 *Al-nujum al-zahira fi muluk Misr wa'l-Qahira*. Cairo.

Al-Tabari, Abu Ja'far Muhammad b. Jarir.
1964 *Ta'rikh al-rusul wa'l-muluk* . Leiden.

Al-Waqidi,
1966 *Kilab al Maghazi*. London 1966.

Secondary Sources

Applebaum, S.
1966 Research on Early Agriculture in the Central Negev. *Yediot – Bulletin of The Israel Exploration Society, Jerusalem* 30:238-224. (Hebrew).

Ariel, D.T.
in press "The Coins from Ashkelon, the 'Third Mile Estate'", in *Excavations in the "Third Mile Estate"*. Ashkelon.

Avner, R.
1995 "Eilot". *Hadashot Arkheologiyot* 103: 105-106. (Hebrew).
1998 "Eilat – Eilot: A Village from the Early Islamic Period". *'Atiqot* XXXVI: 21-39. (Hebrew).

Avni, G.
1985 "Negev Emergency Survey – Har Loz and Har Hame'ara (Maps 206-207)". *Hadashot Arkheologiyot* 86: 40-41. (Hebrew).
1992 *Archaeological Survey of Israel, Map of Har Saggi – Northeast (225) 13-03*. Jerusalem: Israel Antiquities Authority.
1996 *Nomads, Farmers and Town-Dwellers: Pastoralist-Sedentist Interaction in the Negev Highlands. Sixth – Eight Centuries CE.* Publications of the Archaeological Survey of Israel; supplement. Jerusalem: Israel Antiquities Authority.

Baumgarten, Y.
2004 *Archaeological Survey of Israel, Map of Shivta – Northeast (166).* Jerusalem: Israel Antiquities Authority.

in press, a. *Archaeological Survey of Israel, Map of Mashabe Sade (163).* Jerusalem: Israel Antiquities Authority.

in press, b. *Archaeological Survey of Israel, Map of Revivim – Northwest (159).* Jerusalem: Israel Antiquities Authority.

Brock, S.P.

1982 "Syriac Views of Emergent Islamin" in *Studies On the First Century of Islamic Society*, ed. G.H.A. Joynboll. Carbondale, IL: Southern Illinois University. pp 9-21.

Cahen, C.

1995 "Djiza". *Encyclopedia of Islam* II (New Edition): 559. Leiden: Brill.

Cohen, R.

1981 *Archaeological Survey of Israel, Map of Sede Boqer – East (168) 13-03.* Jerusalem: Israel Antiquities Authority.

1985 *Archaeological Survey of Israel, Map of Sede Boqer – West (167) 12-03.* Jerusalem: Israel Antiquities Authority.

1999 *Ancient Settlement of the Central Negev. Volume: The Chalcolithic Period, the Early Bronze Age Period, and the Middle Bronze Age I.* Jerusalem: Israel Antiquities Authority.

Cohen, R. and Cohen-Amin, R.

2004 *Ancient Settlement of the Negev Highland. Volume 2: The Iron Age and the Persian Period.* Jerusalem: Israel Antiquities Authority.

Conrad, L.I.

1986 "The Plague in the Bilad al-Sham in Pre-Islamic Timesin" in *Proceedings of the Symposium on Bilad al-Sham During the Byzantine Period. November 15-19, 1983, II*, eds. M.A. Bakhit and M. Asfour. Amman: Jordan University. pp. 143-163.

Di Segni, L.

1997 Dated Greek Inscriptions from Palestine from the Roman and Byzantine Periods. Vol. I, A: Text. Unpublished PhD Dissertation, The Hebrew University in Jerusalem.

Donner, F.M.

1989 "The Role of Nomads in the Near East in Late Antiquity (400-800 C.E.)" in F.M. Clover and R.S. Humphreys, eds. *Tradition and Innovation in Late Antiquity.* Madison, Wis.: University of Wisconsin. pp. 73-85.

Fabian, P.

1996. "Evidence of Earthquake Destruction in the Archaeological Record – The Case of Ancient Avdat" in *Big Cities World Conference on Natural Disaster Mitigation in Conjunction with the Tenth International Seminar on Earthquake Prognostics*, Abstracts, Jan. 5-10, 1996, Cairo, Egypt. p. 25.

Foss, C.

1997 "Syria in Transition, A.D. 550-750. An Archaeological Approach". *Dumbarton Oak Papers* 51: 189-269.

Gil, M.

1983 *Eretz Israel during the First Muslim Period (634-1099).* Tel Aviv: Publications of the Diaspora Research Institute.

Griffith, S. H.

1993 "A Chapter in the History of the Holy Land According to Early Arabic Texts from Southern Israel". *Cathedra* 69: 57-78.

Haiman, M.

1986 *Archaeological Survey of Israel, Map of Har Hamran – Southwest (198) 10-00.* Jerusalem: Israel Antiquities Authority

1990 "Agricultural Settlement in Ramat Barnea in the 7th – 8th c. C.E." *'Atiqot* 9: 111-124.

1991 *Archaeological Survey of Israel, Map of Mizpe Ramon – Southwest (200).* Jerusalem: Israel Antiquities Authority.

1993 *Archaeological Survey of Israel, Map of Har Hamran – Southeast (199).* Jerusalem: Israel Antiquities Authority.

1995 "Agricultural and Nomad-State Relations in the Negev Desert in the Byzantine and Early Islamic Periods". *Bulletin of the American School of Oriental Research* 297: 29-53.

1999 *Archaeological Survey of Israel, Map of Har Ramon (203).* Jerusalem: Israel Antiquities Authority.

Haldon, J.F.

1990 "Byzantium in the Seventh Century". *The Transformation of Culture.* Cambridge: Cambridge University.

Israel, Y. and Erickson-Gini, T.

in press *Excavations in the "Third Mile Estate".* Ashkelon.

Israel, Y., Nahlieli, D. and Ben-Michael, Y.

1995 "The Nahal Shaqaq Site: A Settlement from the Early Islamic Period in the Northern Arava". *'Atiqot* XXVI: 1*-14*.

Kennedy, H. and Liebeschuetz, H.W.G.

1988 "Antioch and the Villages of Northern Syria in the Fifth and Sixth Centuries A.D.: Trends and Problems". *Nottingham Medieval Studies* 33: 65-90.

Kraemer, C.J.

1958 *Excavations at Nessana III. Non-Literary Papyri.* Princeton: Princeton University.

Lecker, M.

1989 "The Estates of 'Amr b. al-'As in Palestine: Notes on a New Arabic Inscription". *Bulletin of the School of Asian and African Studies* 52: 24-37.

Lender, Y.

1990a *Archaeological Survey of Israel, Map of Har Nafha (196) 12-01.* Jerusalem: Israel Antiquities Authority.

1990b *Ancient Rock Inscriptions. Supplement to Map of Har Nafha (196) 12-02.* Jerusalem: Israel Antiquities Authority.

Lester, A.

1990 "Appendix: Glass Weights from the Reign of Abd el Malik bin Yaziz". *'Atiqot* 10: 125-126.

Magness, J.

2003 *The Archaeology of the Early Islamic Settlement in Palestine.* Winona Lake: Eisenbrauns.

Mayerson, P.

1983 "The City of Elusa in the Literary Sources of the Fourth-Sixth Centuries". *Israel Exploration Journal* 33: 247-253.

Nahlieli, D.
1999 The Negev between the 7th and 11th Centuries CE in Light of Archaeological Findings and Historical Research. Unpublished PhD Dissertation, The Hebrew University in Jerusalem. (Hebrew).
Nahlieli, D., Israel, Y. and Ben-Michael, Y.
1996 "The Nahal La'ana: An Early Islamic Farm in the Negev". *'Atiqot* XXX: 67*-78*.
Nahlieli, D., Israel, Y. and Ben-Michael, Y.
in press *Nahal Omer: A Village from the Early Islamic Period in the Arava.*
Negev, A.
1993 "The Nessana Papyri and the End of the Byzantine Nabataean Settlement in the Negev" in *Studies on the Archaeology and History of Eretz Israel in Honor of Moshe Dothan*, eds M. Heltzer; A. Segal; and D. Kaufman. Haifa: Haifa University. pp. 231-42.
Palmer, A.
1993 *The Seventh Century in the West Syrian Chronicles*. Liverpool: Liverpool University.
Porat, Y.
1987 "'Tunnel Wells' in the Arava" in *Eilat and the Arava*, eds Abiram *et al.* Jerusalem: Publisher. (Hebrew). pp. 243-60.
Rosen, S.
1985 "Negev Emergency Survey – Mizpe Ramon (Map 201)". *Hadashot Arkheologiyot* 86: 40. (Hebrew).
1994 *Archaeological Survey of Israel, Map of Maktesh Ramon (204).* Jerusalem: Israel Antiquities Authority.
Sharon, M.
1976 "The Process of Destruction and Nomadization in Eretz Israel under Islamic Rule (633-1517 C.E.)" in *Chapters in the History of Eretz Israel under Islamic Rule*, ed. M. Sharon. Jerusalem: Yad Yizhaq Ben-Zvi (Hebrew). pp. 9-32.
1977 "The Beduin and Eretz Israel under Islamic Rule". *Zohar*: 199-210 (Hebrew).
1981 "Appendix: Arabic Inscriptions from Sede Boqer" (Hebrew), xxxi*-xxxii*(English) in *Archaeological Survey of Israel, Map of Sede Boqer – East (168) 13-03, R. Cohen.* Jerusalem: Israel Antiquities Authority. pp. 87-88.
1985 "Appendix: Arabic Inscriptions from Sede Boqer – Map 167" in *Archaeological Survey of Israel, Map of Sede Boqer – West (167) 12-03*, R. Cohen. Jerusalem: Israel Antiquities Authority. pp. 90-106 (Hebrew), 31*-35* (English).
1990 "Arabic Rock Inscriptions from the Negev" in *Ancient Rock Inscriptions. Supplement to Map of Har Nafha (196) 12-02*, Y. Lender. Jerusalem: Israel Antiquities Authority. pp. 9-36 (Hebrew), 9*-35* (English)
1993 "Five Arabic Inscriptions from Rehovoth and Sinai". *Israel Exploration Journal* 43: 50-59.
Tsafrir, Y.
1996 "Some Notes on the Settlement and Demography of Palestine in the Byzantine Period: The Archaeological Evidence" in *Retrieving the Past. Essays on Archaeology and Methodology in Honor of G.W. Van Beek,* ed. J.D. Seger. Michigan: Cobb Institute of Archaeology, Michigan State University. pp. 269-283.
Tsafrir, Y. ed.
1988 "Excavations at Rehovot-in-the-Negev". *Qedem* 25. Jerusalem: Hebrew University.
Ustinova, Y. and Nahshoni, P.
1994 "Salvage Excavations in Ramot Nof, Be'er Sheva". *'Atiqot* XXV: 157-177.
Zohar, A.
1977 *The Desert: Past, Present and Future.* Tel Aviv: Reshafim (Hebrew).

The Other Side of the Coin: The use of Milled Colonial Spanish Coins as Medicinal Talisman Among the Bedouin and Fellahin in southern Palestine

Benjamin Adam Saidel and Abd Barakat

Introduction

A truism of archaeological and ethnographic research is that mobile pastoralists are integrated into, and dependent upon, symbiotic relations with the "Outside World" (e.g., Barfield 1993: 16, 68-69, Khanazov 1994; Marx in this volume; Musil 1928: 278-279; Rosen 2000, 2002: 35-36). In the 19th century, the Bedouin in Arabia, the Levant and Sinai acquired Milled Spanish Silver coins through their economic activities. Aside from functioning simply as a medium of exchange, however, the symbols on the reverse side of these coins were reinterpreted and contextualized within the culture of Bedouin and Fellahin societies. As a result, this coin came to be used as a medicinal talisman by these populations during the Ottoman and British Mandate periods.

The use of Milled Colonial Spanish coins in the folk traditions of Bedouin and Fellahin societies was discovered by one of the authors (AB) in the course of his long-term commercial interactions with these communities.[1] This paper describes how the Bedouin acquired Milled Colonial Spanish coins, and how the symbols on the reverse of these coins were integrated into the folk culture of Bedouin and Fellahin societies of southern Palestine during the 19th and early 20th centuries C.E.

Milled Spanish silver coins

Milled Spanish silver coins were struck by the Spanish Empire at a number of mints in Spain and in Central and South America between 1732 and 1821 (Jordan 1997). Prior to 1772, the obverse of these silver coins depicted the crowned arms of Castile and Leon, while the reverse side contained two orbs flanked to the left and right by the pillars of Hercules. After 1772, the busts of Spanish kings replaced the crowned arms of Castile and Leon on the obverse, and on the reverse the pillar-flanked orbs were replaced by the crowned shield of Castile and Leon, now flanked by the two pillars (Fig. 1). These changes in design were carried out in order to distinguish earlier silver issues from later ones, as the latter coins contained less silver than the former (Jordan 1997).

Fig. 1. A 8 Real-pillar coin, formerly from the ethnographic collection of Yacub Harub (Jerusalem), used by Bedouin and/or Fellahin as a talisman. This coin was minted in Mexico City either in 1790 or 1803. Obverse: Bust of Charles IV of Spain with inscription: Carolus IIII ...Gratia...; Reverse: Crowned shield of Castile and Leon flanked by Pillars of Hercules with inscription: Hispan et Ind Rex Mo R.

For the Kingdom of Spain, Milled Spanish Colonial Coins were both a currency and a propaganda vehicle for displaying the wealth and power of Spain and its colonies.[2] The two pillars were meant to represent the Pillars of Hercules, which flanked the straights of Gibraltar and symbolized the end of the known world according to Classical authors (Jordan 1997). In addition, a banner, which wound around the two pillars on the reverse of the coin, was stamped with the Latin phrase "PLUS ULTRA," translated as "more beyond" and was meant to refer to the Spanish colonial conquests and newly acquired wealth in the Americas (Jordan 1997).

[1] Abd Barakat is a collector of ethnographic material culture from Bedouin and Fellahin societies.

[2] On the use of coinage for propaganda purposes see Grierson (1975: 76).

Spanish Pillar coins entered the Levant due to the export of cotton from Acre to Europe during the 18th century (Kool 2005). These coins, and later the Austrian Maria Theresa thalers, supplanted the silver grush and/or piaster issued by the Ottoman Empire (Kool 2005). Massive quantities of Milled Colonial Spanish silver coins entered the Middle East via trade between Europe/Spain and Egypt from 1690 to 1720. During this period between 300,000 and one million Spanish dollars were sent from the port of Marseille to Egypt as payment for coffee and textiles (Walz 1983: 307). By 1730, however, Spanish dollars continued to be imported into Egypt, but in smaller quantities (Walz 1983). However, by the beginning of the 19th century this type of coin was the preferred silver currency among Bedouin in Arabia (Burckhardt 1829: 43). For example, when Mohammed Ali Pasha attempted to remove the Spanish Dollar from circulation in Arabia and supplant it with Egyptian *piastres* the Bedouin baulked (Burckhardt 1829: 43):

> "In the Hedjaz [Mohammed Ali Pasha] has not the same means of enforcing his despotic measures to their full extent; and thus it happens that in the interior of the country, where the Turkish troops are placed, the value of the dollar is eighteen or nineteen piastres. The Bedouins, however, refuse to take the Egyptian piastres, even at a depreciation, and will receive nothing but dollars; a determination to which the Pasha himself has been frequently obliged to yield."

By the time of Burton's travels in Arabia (1853-1854), milled Spanish coins were no longer being minted by Spain and as a result they were being replaced by increasing issues of Maria Theresa Thalers. Nevertheless, Spanish dollars were one among many coins in circulation in Arabia when Burton visited this country (1865/1964: 111, footnote 1):

> "The Spanish dollar is most prized in Al-Hijaz; in Al-Yaman the Maria Theresa. The Spanish Government has refused to perpetuate its Pillar-dollar, which at one time was so great a favourite in the East. The traveller wonders how "Maria Theresas" still supply both shores of the Red Sea. The marvel is easily explained: the Austrians receive silver at Milan, and stamp it for a certain percentage. This coin was doubtless preferred by the Badawin for its superiority to the currency of the day: they make from it ornaments for their women and decorations for their weapons. The generic term for dollars is "Riyal Fransah."

By the time of Doughty's travels in Arabia milled Colonial Spanish coins were no longer issued by the Government of Spain, but they remained in circulation among merchants in Arabia (Doughty 1888/1979: 23):

> "All their dealings are in foreign money; reals of Spain, Maria Theresa dollars and Turkish *mejîdy* crowns; gold money is known more than seen among them. They call *doubloon* the piece of 5 Turkish pounds, English sovereigns *ginniyát* or *bintu*, and the 20 fr. piece *lira fransâwy*."

Both Burton and Doughty noted that trade coins circulating in Arabia were generally referred to as "riyals" or "reals," terms originally used solely for the Spanish dollar (Burton 1865/1964: 111, footnote 1; Doughty 1888/1979: 664).[3]

Milled colonial Spanish coins were acquired by the Bedouin through a range of commercial activities. Doughty observed that Spanish dollars functioned as a basic medium of exchange for the Bedouin in Arabia as they "reckoned only by reals" (Doughty 1888/1979: 430). He also attributed the presence of milled colonial Spanish coins to commerce carried out in ports located on the Persian Gulf (Doughty 1888/1979: 664). The Bedouin also acquired a range of silver coinage through the "sale of some of their camels to the brokers" (Doughty 1888/1979: 338). Doughty notes that one source of silver currency, the Turkish mejîdies, was paid to those Bedouin sheikhs who lived on or adjacent to the Haj road (Doughty 1888/1979: 338). In addition, travelers such as Doughty used Spanish dollars in their commercial transactions with the Bedouin (Doughty 1888/1979: 63).

Abu 'Amud

Bedouin and Fellahin referred to Milled Silver Spanish Coins as Abu 'Amud, which can be literally translated as "Father of the Pillar" (Weir 1989, 176, footnote 16; Meir 2002: 14*), a name derived from the dual pillar symbol on the reverse side of the coin. On western coinage it is typical for the obverse to carry "the more important device or inscription," which functions to denote "the authority by which the coin is issued" (Grierson 1975: 86, 195). The name "Abu 'Amud," however, suggests that, for the Bedouin and Fellahin, the significance of the coin and its symbols lay in the reverse.[4] Issues of these 'Pillar' coins from 1732 to 1821 had the symbols of the Spanish monarchy on the obverse, representing the authority and institution of the Spanish crown supporting the value of the currency. For the Bedouin and Fellahin in Arabia and the southern Levant these markers of the Spanish monarchy, symbols such as the banner inscription, the crowned arms of Castile and Leon, and the bust of specific Spanish Kings, appear not to have had much relevance as references and suggests that they either did not recognize the images as referring to the "Pillars of Hercules," and the inherent authority vested in this symbol or alternatively chose to promote a local referent, one that more clearly resonated within their local circles.

---

[3] It is interesting to note that the currency in the modern country of Saudi Arabia is called the *riyal* (Ross 1994: 41).

[4] Spanish Milled Silver coins were not the only coins identified on the basis of the designs on their reverse. For example, Thalers, including the Maria Theresa Thaler, were usually identified by the eagle on the reverse of these coins, hence this type of coin was generally referred to as *Abu Risheh*, "father of feathers" (Weir 1989: 176, footnote 16; Meir 2002: 14).

Alternatively, the two Pillars of Hercules on the reverse were images which were easily incorporated as meaningful symbols into Bedouin and Fellahin traditional folk culture. There seem to be two possible interpretations for the Pillars. The more popular explanation is that the pillars were viewed as phallic symbols, an interpretation which influenced the role that this coin played in folk medicine. A more speculative suggestion is that the pillar image resembled a type of ancient monument found in the arid zones of the southern Levant and Arabia. According to Burton (1865/1964: 109) the Bedouin sometimes referred to the Spanish dollar as follows:

> "The Badawi calls a sound dollar 'Kirsh Hajar,' or 'Riyal Hajar,' a 'stone dollar'."

It may be possible to suggest that the Bedouin drew analogies between the shape of the Pillars of Hercules and the standing stones or menhirs, referred to as *massebot* in Israel, found in Arabia, Sinai and in the arid zones of the southern Levant (e.g., Avner 1984: 115-119; Zarins 1992: 51).[5]

Medicinal Talismen

Information collected by one of the authors (AB) suggests that the 'Abu Amud' coin was used, not only as currency, but as a medical talisman by both males and females (Fig. 2). As a talisman in folk medicine, Spanish Milled coins could either be used solely or attached to beads.[6] As a prophylaxis from illness young boys could hold the coin on their head and have a family member pour water over the coin in order to ward off the evil eye. Among women the Abu Amud coin was used for a variety of rituals associated with fertility and pollution. For example, when a woman visits another woman who has recently given birth, both have to wash underneath the coin. The reason stated is that a new mother is considered 'polluted' and therefore the visitor needs to be cleansed and protected so that her own fertility will not be jeopardized. In another context, following her period, a woman might wash underneath this type of coin as a method for increasing her fertility. We discovered at least two other variations of ritual cleansing practice. In one account a woman might stand on this coin while she bathes, or the coin might be placed underneath her pillow while she sleeps.

The Bedouin's use of the Abu Amud coin as a medical amulet in the Ottoman and Mandate periods was most likely based on practical considerations. Archaeological excavations of the Bedouin cemetery at Tell el-Hesi, dated from the 15th to 19th centuries A.D., demonstrate a high rate of both infant mortality and death of young adult

Fig. 2. Two girls from the Ta'amreh tribe photographed in the early 20th century, when the tribe lived to the east of the village of Artas. The older girl (left) has a silver trade coin (similar to a Maria Theresa thaler and a Spanish Pillar coin) attached to her dowry hat. Her dowry hat with the attached row of silver coins indicates that she is already spoken for in marriage. Her clothing is made from pieces of old dresses with good-luck charms. The talisman and the fragments of dresses originate in areas of Hebron and Yatta coins (From Studio Varouj).

women most likely associated with the dangers of childbirth among Bedouin (Eakins 1993: 79). Writing at the beginning of the 20th century Jennings-Bramley observed that the Bedouin in the Sinai did not have access to "doctors of any kind or nationality" (Jennings-Bramley 1905: 132). He further noted that if one was available he would most likely be stationed only at the town of El Arish, in Sinai. Furthermore, Jennings-Bramley believed that the Bedouin were often reluctant to pay for medical services (Jennings-Bramley 1905: 133).[7] During the British Mandate the Bedouin of the Negev had access to limited medical facilities in Beersheva. According to Aref el Aref, who was governor of the southern district of Palestine from 1929 to 1939 and based in Beersheva, there was "only one Government doctor and only one licensed practitioner" in the entire subdistrict of Beersheva (el-Aref 1944: 69). Moreover, those Bedouin who lived to the east and south of the town of Beersheva tended to be distrustful of modern medicine (Abu-Rabi'a 2001: 52).

---

[5] For an example of standing stones/*massebot* depicted on the reverse side of a Hellenistic coin, see the example from Phoencia which is illustrated in Avner (1993: 172, Fig. 22; 174-175).

[6] For examples of other types of coins used as medicinal amulets see Meir (2002:11*, 19, Figure 15, 20, Figure 16) and Helmecke (1998: 47, Nos. 133, 134, 48, No.143).

[7] In the Kerak area of Transjordan, the Bedouin sought medical services from Forder, who was both a doctor and missionary. Most cases that are mentioned by Forder are injuries that the Bedouin sustained while fighting (e.g., Forder 1909/2002: 36, 48-49), and not those associated with childbirth.

Tawfik Canaan, a physican and ethnographer, demonstrated that by 1925 high infant mortality rates were present among both 'peasants' and 'poor city dwellers' in Palestine. Out of 2,185 medical cases he documented 20% of the infants died at birth and another 36% died following birth. In the course of his medical practice Canaan noted that gold coins circulating in Ottoman Palestine were used as talisman and as medical amulets to treat a range of maladies (Canaan 1914: 114). Canaan collected a number of these coins. Examples from his personal collection include Byzantine and Ottoman coins which were used to promote "healing from diseases" (Helmecke 1998: 47, 133, 134).[8]

The appeal of milled colonial Spanish coins as amulets in traditional folk medicine increased due to the increasing scarcity of this coin in the early part of the 20th century. At that time it was estimated that the ratio of Spanish Milled Coins to Maria Theresa Thaler coins was approximately 1 to 100. While Spanish Milled Coins were minted between 1732 and 1821, Maria-Theresa Thalers continued to be minted well into the 19th century for export to the Near East even after they went out of circulation in Austria.[9] As a result of the rarity of Abu 'Amud coins only a few families, such as a village Mukhtar or similar figure, actually owned an 'Abu 'Amud'. The coin was then rented to those who needed to use it as a medicinal amulet. Therefore, it was not uncommon for an 'Abu 'Amud' coin to move between families.

Conclusion

Often archaeological and ethnographic research are cited as evidence that mobile pastoralists are economically integrated into the "Outside World" (e.g., Barfield 1993: 16, 68-69, Khanazov 1994; Marx in this volume; Musil 1928: 278-279; Rosen 2000, 2002: 35-36). Milled Spanish coins are an example of Western material culture which was integrated into the folk medicine of Bedouin and Fellahin in southern Palestine during the Ottoman and Mandate periods. These populations, however, did not interpret the Pillars of Hercules as symbols of Spanish conquest, but rather as symbols of strength and virility which could pass from the coin to a person once it was imbued with the proper ritual. The presence of these coins in Bedouin communities corresponds with a growing line of evidence, such as coffee, firearms, and tobacco, which demonstrate that the Bedouin in the Ottoman and Mandate periods were integrated into a global economy (e.g., Baram 1996; Saidel 2000).

[8] Canaan acquired these three amulets at different times. The amulet containing a silver Byzantine coin was acquired in Jerusalem in 1929, and the talisman containing a gold Byzantine coin was collected in 1942 (Helmecke 1998: 47, 133, 134). The third amulet, which contained an Ottoman coin among other items, was purchased in Lifta in 1913 (Helmecke 1998: 48, 143).

[9] In western Arabia and Yemen Maria-Theresa Thalers were used as currency until the 1930s (Great Britain Naval 1945: 537).

Acknowledgments

We wish to thank Laura Mazow and Eveline van der Steen for their comments on earlier drafts of this paper.

Bibliography

Abu-Rabi'a, A.
2001 *A Bedouin Century: Education and Development among the Negev Tribes in the 20th Century*. New York: Berghahn Books.
El-Aref, A.
1944 *Bedouin Love, Law, and Legend: Dealing Exclusively with the Badu of Beersheva.* Jerusalem: Cosmos Publishing Company
Avner, U.
1984 "Ancient Cult Sites in the Negev and Sinai Deserts." *Tel Aviv* 11: 115-131.
Avner, U.
1993 "Mazzebot Sites in the Negev and Sinai and their Significance" in Biran, A. and Aviram, J. (eds.), Biblical Archaeology Today, 1990. Proceeding of the Second International Congress on Biblical Archaeology, Jerusalem, June-July. pp. 166-181.
1990 *Jerusalem*: Israel Exploration Society and Israel Academy of Sciences and Humanities.
Baram, U.
1996 *Material Culture, Commodities, and Consumption in Palestine, 1500-1900.* Amherst: University of Massachusetts Amherst.
Barfield, T. J.
1993 *The Nomadic Alternative*. Englewood Cliffs, New Jersey, Prentice Hall.
Burckhardt, J. L.
1829 *Travels in Arabia.* London: Association for Promoting the Discovery of the Interior of Africa.
Burton, R. F.
1865 *Personal Narrative of a Pilgrimage to Al-Madinah & Meccah.* Reprinted New York: Dover Publications, Inc, 1964.
Canaan, T.
1914 *Aberglaube und volksmedizin im lande der Bibel.* Hamburg : L. Friederichsen & co.
Doughty, C. M.
1888 *Travels in Arabia Deserta.* New York, Dover Publications Inc., 1979
Eakins, K. J.
1993 *Tell el-Hesi: The Muslim Cemetery in Fields V and VI/IX (Stratum II).* Winona Lake, Indiana: Eisenbrauns.
Forder, A.
1909 *Ventures Among the Arabs in Desert, Tent, and Town.* New York: Gospel Publishing House. Reprinted Gorgias Press, 2002.
Great Britain
1946 *Western Arabia and the Red Sea.* Geographical Handbook Series B.R. 527. Oxford: Naval Intelligence Division.
Grierson, P.
1975 *Numismatics*. Oxford: Oxford University Press.
Helmecke, G.

1998 "The Tawfik Canaan Collection of Palestinian Amulets" in K. Nashef (ed.), Ya kafi, ya shafi... The Tawfik Canaan Collection of Palestinian Amulets: An Exhibition, October 30, 1998-February 25, 1999. Birzeit University. pp. 27-50.

Jennings-Bramley, W. E.

1905 "The Bedouin of the Sinaitic Peninsula". *Palestine Exploration Quarterly* 126-137.

Jordan, L.

1997 Colonial Milled (Pillar) Coinage 1732-1821: Introduction. http://www.coins.nd.edu/ColCoin/ColCoinIntros/Sp-milled.intro.html (last accessed 12-12-2005)

Khazanov, A. M.

1994 *Nomads and the Outside World.* Madison, University of Wisconsin Press.

Kool, R.

2005. *The Circulation of Dutch Leeuwendaalders in 17th Century Ottoman Palestine: New Evidence from Two Unpublished Hoards from the IAA.* http://www.antiquities.org.il/article_Item_eng.asp?sec_id=17&sub_subj_id=374#MMMas (last accessed 12-13-2005)

Meir, C.

2002 *Crown of Coins. Traditional Headdresses of Arab and Bedouin Women.* Tel Aviv: Eretz Israel Museum.

Musil, A.

1928 *The Manners and Customs of the Rwala Bedouins.* New York, American Geographical Society.

Rosen, S. A.

2000 "The Decline of Desert Agriculture: A View from the Classical Period Negev" *The Archaeology of Drylands: Living at the Margin.* G. Barker and D. Gilbertson. London and New York, Routledge. 39.

Rosen S.A.

2002 "The evolution of pastoral nomadic systems in the southern Levantine periphery" in *In quest of ancient settlements and landscapes*, eds. E.C.M. van den Brink and E. Yannai. Tel Aviv: Ramot Publishing. pp. 23-44.

Ross, H.C.

1994 The Art of Bedouin Jewelry: A Saudi Arabian Profile. 4th Edition. Studio City: Empire Publishing Service/Players Press.

Saidel, B. A.

2000 "Matchlocks, Flintlocks, and Saltpetre: The Chronological Implications for the Types of Firearms used by the Bedouin during the Ottoman period." *International Journal for Historical Archaeology* 4(3): 191-216.

Walz, T.

1983 "Gold and Silver exchanges between Egypt and Sudan, 16th-18th Centuries" in Richards, J.F. (ed.), Precious Metals in Later Medieval and Early Modern Worlds. Durham: Carolina Academic Press. pp. 305-325.

Weir, S.

1989 *Palestinian Costume.* Austin: University of Texas Press.

Zarins, J.

1992 "Archaeological and Chronological Problems within the Greater Southwest Asian Arid Zone" in Ehrich, R (ed.), *Chronologies in Old World Archaeology.* Chicago: University of Chicago Press. pp 42-76.

Town and Countryside of the Kerak Plateau

Eveline J. van der Steen

Introduction: The Tribal Society

In archaeology there has always been a tendency to create an 'archaeology of the town' and an 'archaeology of the countryside'. A number of different theories have been developed, ranging from models that stressed the opposing positions of town (or city) and countryside (Joffe 1993:90 states: Canaanite urbanism ultimately existed in opposition to other social forms, which resisted and rebelled against urban domination and exploitation), to models that stressed the interaction between the two (see Herzog 1997:1-13 for an extensive overview of the different approaches; and most recently the critique of the 'two sector model' in Schloen 2001:221-254).

Even in the models that stressed interaction, the tendency was to separate the town-dwelling population from that of the countryside. According to Joffe there was, starting in the Early Bronze II, a clear separation between the 'urban dweller' and the 'rural dweller' or 'nomad' (Joffe 1993:72). It is generally assumed that these two social groups were also differently organized internally. The urban population is usually depicted as being organized in different social classes headed by an administrative and/or religious elite, that was responsible for the organization of collective projects, the redistribution of goods and generally the running of the town. The countryside, on the other hand, was organized in a more egalitarian way, shared by village-based farmers and nomadic pastoralists. Much has been written about the way these three segments of society interacted, and numerous models have been developed. Rowton (1973a, 1973b, 1974) developed the model of the 'dimorphic society', describing the interaction between nomadic, rural and urban groups in or on the fringe of a sedentary environment. His model has been criticized and supplemented by other models, such as that of the 'polymorphous' or 'multimorphic' society (Salzman 1980:1-19; Finkelstein 1995:26), which states that there is a continuum in economic pursuits that are (at least theoretically) accessible by the whole population of a certain society, so that people or groups of people can move along from one economic pursuit to another when that is economically profitable or desirable.

In spite of the fact that tribes are prominent in the written sources from the Bronze Age on, they seldom play a role in archaeological interpretation in the southern Levant. There have been efforts to relate the origins of Israel in the Early Iron Age to the Egyptian literary sources that mention Shasu (recently Rainey 2001, 2003, with references), interpreting the Shasu as a tribe, or a tribal confederation. However, in general tribes remain elusive in archaeology. One reason for that is that tribes are generally associated with a nomadic lifestyle, comparable to that of the later Bedouin. It is not unusual in archaeology to talk about 'bedouin tribes' in an early Iron Age or even earlier context, for example.

Recently the picture of the town as a class-based community has been challenged. Philip (2001) has suggested that, instead of having a hierarchical, class-based organization, towns and villages in the Early Bronze Age may have been inhabited by corporate groups, that cooperated in major collective undertakings such as monumental architecture, walls, temples or palaces. The same applies, obviously, to later periods, and perhaps earlier ones as well.

Chesson (2003) has proposed a social organization based on the concept of the "house society" which explains how this society may have functioned on the level of the separate sections or family groups.

On a different scale, LaBianca (1999) has introduced and described the concept of the Tribal Kingdom as a heterarchical form of social organization in the Iron Age, consisting of interacting tribes, challenging the traditional vertically organized state structure and replacing it with a horizontal one. One of the features he addresses is the relation between town and countryside in a Tribal Kingdom (LaBianca 1999:21). In LaBianca's model the tribal hinterland is administered from a fortified town, containing a cluster of administrative buildings with a 'cadre' of bureaucrats administering the economic affairs of the actual population, who live mainly in the hinterland. This model has been applied to Iron Age Edom, which has been interpreted as a tribal kingdom, consisting of interacting tribes, in a heterarchical power balance, sharing an economy of pastoralism, agriculture, involvement in the Arabian trade, and copper mining (Bienkowski and van der Steen 2001).

Case Studies

Building on LaBianca's concept of tribal organization as the basis for a heterarchical, multi-level society, and using etnoarchaeological sources, I want to take a closer look at the interaction between town and countryside in a tribally organized society, using the town of Kerak in the 17th-19th century AD as a primary case study. Other case studies are provided by two other towns, Salt and Jauf, in order to put the case of Kerak in perspective, and arrive at more general observations.

Kerak: short history

The later part of the Ottoman period we know rather well, thanks to the accounts of a number of travellers in the region from the end of the 18th century on. Many of these travelled on the Kerak Plateau, and stayed in Kerak. It is these sources in particular that we rely upon for the local history of the region, for which few alternative sources exist (Burckhardt 1822:377-395; Seetzen 1854/I: 420-430; II:328-336; Dissard 1905; Musil 1907-08/I: 25-34; III:70-129; Oppenheim 1943:255-276; Peake 1958:183-191; Gubser 1973:12-25, with references, and many others).

The 17th-19th century AD is a period in which the tribal power structures were fully functioning on the Kerak Plateau. Formally it fell under the authority of the Ottoman empire, but for all practical purposes the control of the empire did not extend south of Jerusalem in the west, and south of Salt in the east (Burckhardt 1822:265; Gubser 1973:14; Lewis 1987:23). This does not mean that there were no contacts with the Ottoman empire: there were occasional battles between the Turkish army and the bedouin, or efforts (generally fruitless) by the government to collect taxes; moreover, the government paid protection money to the bedouin in order to protect the Pilgrim's route to Mecca, and occasionally traded with them, especially in the gateway towns.

The stronghold of Kerak had been created by the Crusaders in the 12th century, and fortified by the Mamluks. When these external powers lost interest in the area in the 16th century, it became part of the tribal society of the Plateau. Until the 18th century the tribes that controlled the Plateau also controlled the town. In the beginning of the 18th century the Plateau was controlled by the Beni 'Amr, who had divided the land among their different sections. Kerak became part of the territory of one of these sections: the Ibn Qaisum, who used the town mainly to extract taxes from.

A section of the Ibn Qaisum lived in the town, where the sheikh had a residence, although he preferred to spend the winter in the Ghor. The Beni 'Amr were traditionally pastoralists, and they had come from the Hijaz as camel breeders. Once they had conquered their new territory on the Kerak Plateau, various sections of the Beni 'Amr adapted their economy to suit their new surroundings: the Ibn Tebet in the east continued to breed camels, but the economy of the Ibn Qaisum, who controlled the western part of the Plateau, was based on agriculture, animal husbandry, as well as robbing and extorting protection money from other, minor tribes and from the town and villages.

Inside the town there lived a number of tribes and sections of tribes. One of these, the Imamiyeh, had been the leading tribe of the town from the 16th century on, although they were subjected to the Ibn Qaisum. At the beginning of the 18th century the Majali, descendants of a Hebron merchant family, who lived in the town, allied themselves with the Beni Amr, and defeated the Imamiyeh. The Majali sheikhs were clever politicians. After they had beaten the Imamiyeh, they managed to raise one section of the Beni Amr against another, and eventually beat them with the help of the Beni Sakhr, the Hameide and the Hajaja, other tribes that lived on the plateau. By the end of the century the Majali controlled not only the town, but also the Plateau. In the following century they were almost continually involved in territorial struggles and raids with the major tribes of the region, including their former allies: the Beni Sakhr, the Howeitat, the Hajaja and the Hameide. Eventually the conflict with the Beni Sakhr led to the end of the independence of the Kerak Plateau, when the Beni Sakhr called in the help of the Ottoman government in 1893.

Social organization

The social and spatial organization on the Kerak Plateau was reflected in the social and spatial organization of the town. The tribes on the Plateau were organized in two major alliances: the Eastern and the Western alliance (Gubser 1973: map 4). The western alliance was headed by the Majali, the eastern by the Tarawneh. The christian tribes generally sided with the western alliance. Most tribes had sections living both on the Plateau and in the town (Burckhardt 1822:377,388; 1831:16; Musil 1908:76; Jaussen 1908:71; Gubser 1973:51). In the town itself the eastern and western alliances had their own section, as had the christians. Within this major division each tribal section or family had its own quarter. This spatial organization was still partly reflected in the actual lay-out of the town up to the 1960s (Gubser 1973: map at the end). The leading tribe of the town, the Majali, was also the leading tribe of the Plateau. There were also tribes or sections that did not participate in the territorial division of the plateau, largely because they were specialists: craftsmen or traders. They, like the other tribes, each had their own quarter in the town.

The town was the meeting place of the tribes, their market, the place where they interacted, socially and economically, partly through their representatives, and where they created and strengthened economic, commercial and social bonds. The suq was the trade centre in times of peace, and in times of war or tension the neutral area, where no violence should occur (Gubser 1973:57). The specialists provided services that were not available in the countryside, such as certain craft products, and trading facilities. The town was both a reflection of and a central place for the tribal society of the plateau.

The main economic activities of the inhabitants of Kerak were agriculture, horticulture and pastoralism. Many Kerakiyeh camped outside the town on the Plateau during part of the year, to pasture their animals and cultivate their fields. Every 19th century traveller who travelled on the Kerak Plateau mentions coming across camps of the Kerakiyeh, tending their flocks or cultivating the land. Agricultural surplus was bartered with the Hebroni traders in the town, or sold to the Bedouin, or twice a year to the Hajj pilgrims. In fact, traders from Ma'an purchased grain from the Kerak Plateau, in order to to sell it to the Hajj pilgrims (Seetzen 1854:421).

Before the rise to power of the Majali, the people of the town paid for grazing rights and for the right to cultivate on the Plateau, to the Beni 'Amr and other Plateau tribes. After the Beni 'Amr had been evicted, grazing and cultivation rights on the plateau were shared by the tribes of the town, and tribes such as the Hameide and the remaining Beni 'Amr had to pay grazing and cultivation rights to the Kerakiyeh, and particularly to the Majali, the leading tribe (Burckhardt 1822:389; Musil 1908:293). The Plateau was the territory of the Kerakiyeh, divided among the main alliances, and subdivided according to the needs of the separate sections and families (Musil 1908/III:87-89, although Gubser (1973:26) states that there was no *musha* system). The Kerakiyeh, in turn, paid protection fees to the Howeitat and the Beni Sakhr, to keep them from raiding.

This short look at the divisions of power and the power structures on the plateau makes one thing clear: the power of Kerak and the Kerakiyeh was largely due to the leading capacities of one tribe, the Majali, and this was the basis for the power structure in the town as well as on the plateau. It is clear that we cannot build a model on the success of one town, or one family. We need to look at other places to see if we can find general rules. We need to ask the question to what extent the social and economic organization of the Kerak Plateau is representative of the social and economic organization in the region in other periods, in other words, whether we can use it to construct a model for interpretation of the archaeologal record.

Salt

The town of Salt has been visited by a number of travellers in the 19th century, and its history has been extensively described by Abujaber on the basis of family archives and 19th century sources (Abujaber 1989, with references). Salt was the focus of three population groups (Abujaber 1989:69-70). The Saltiyeh, inhabitants of the town itself, consisted of four tribes, each with their own quarters: three muslim tribes and one christian tribe. They were farmers and herders with territory outside the town, but they were also active traders, and they acted as middlemen for the trade between the bedouin and the districts of Jerusalem and Nablus. Next there were the Abbadi, who lived immediately around the town, and were economically dependent on it, for barter and trade of their surplus. Finally there were the Adwan, the paramount tribe of northwestern Transjordan.

Sometime prior to 1800 a land division was effected, in which control of the Belqa lands was divided among these three groups, based on the existing, traditional territorial rights of each group. All three groups were involved in agriculture and pastoralism, but trade was mainly in the hands of the Saltiyeh, the inhabitants of the town. One Saltiyeh tribe, the Haddadin, did not share in the general division of the land, because they were specialists: blacksmiths.

The Saltiyeh had a somewhat ambivalent relationship with the Beni Sakhr, who, on the one hand, were dependent on the town of Salt as a market place, but who on the other hand demanded khawa from the inhabitants, for 'protection' of the Belqa lands (Burckhardt 1822:348,352). Burckhardt's description of the Saltiyeh makes it perfectly clear that they were not subjected to the bedouin tribes of the region, but shared in the control of power and economic assets on an equal basis (Burckhardt 1822:350-1, 367; also Seetzen 1854-I: 402, II:329).

Jauf

A third example is provided by the town of Jauf, in the desert of Saudi Arabia. Jauf was a major oasis, in disputed territory between the Rwala and the Shammar. When Musil came to Jauf in 1908, the Rwala had just conquered the town on the Shammar, with the consent of its inhabitants (Musil 1927:160).

According to Burckhardt (1831:14) the tribe of the Sirhan fled from the Hauran to Jauf in the 17th century. According to various sources the Sirhan remained in Jauf for a period of time ranging from 7 to 20 years. Later most of the Sirhan returned to the Hauran, however, one section remained in Jauf. The returning Sirhan allied themselves to the Beni Sakhr. Wallin (1979:21-44) visited Jauf in 1845. He noted that the inhabitants of Jauf consisted of sections of the Rwala, Shammar and Adwan, as well as immigrants from towns such as Tafileh. Each group lived in their own quarters. Its economy was based on cattle breeding and date growing, and some craft products, such as specially woven mantles and stone coffee mortars, but in the past it had been a major market town, and a trading basis for the Sherarat, who lived around the town. Sometimes the inhabitants would go on raids with the Shammar. Some of the Rwala Bedouin lived in Jauf in the summer, and migrated to the Hauran in winter. Jauf was their point of contact, their market, where economic and social bonds between the tribes were maintained and strengthened. Euting, who visited the town fourty years later, confirms Wallin's observations, and adds that the various sections of the town were continually fighting each other until the Ibn Rashid conquered the town and took control (Euting 1885:124-127).

Discussion: the archaeology of town and countryside

As stated above, prevailing theories about town and countryside see them as two separate entities, sometimes opposed to each other, sometimes interacting, but with differing functions and social and economic organizations. We use terms like nomadic, rural or urban society to distinguish the countryside from the town.

However, the examples mentioned here - and there are many others – suggest that towns in the past can be looked at in a different way, as an integral part of the tribal organization that functioned in the region. The question remains, whether, and how, this social organization is reflected in the archaeological record. There are several possible roads that can be pursued.

First of all, in subrecent societies, tribal organization is expressed in the internal division of the towns. The different tribes, or sections of tribes had quarters in the town, in which they lived at least part of the year. This organization is still reflected in the plans of older towns, and even in more recent plans, such as that of 1960 Kerak. It has also been suggested for several Early Bronze Age towns in Jordan, such as Numeira. Here, according to the excavator, the entire town centre was a domestic area, organized in separate complexes, that, however, shared walls, and was suggestive of systematic town planning along family lines (Coogan 1984:76). Khirbet Zeraqon (Mittmann 1994) had an upper and a lower town (see plan in Philip 2001:173). The public area was adjacent to the town gate, a not uncommon arrangement in Early Bronze Age (and later) towns. Through it one reached the domestic area, which consisted of housing units divided by streets. The whole lay-out is reminiscent of the organization of Kerak in the 19th and early 20th century: clearly defined domestic units, inhabited by corporate groups that may have been individual families or sections of tribes, and a public area, that served as a centre for communication and trade in times of peace, and as a 'neutral' buffer space in times of stress. Greenberg (2003:19) has suggested a similar function for the cultic area in Early Bronze I Megiddo (also van der Steen 2005).

The EB-MB (or Early Bronze IV, or Intermediate Bronze Age) strata of Iktanu, Umm Hammad and Khirbet Iskander show evidence of town planning, in which a number of different 'insulae' could be discerned, that could be interpreted as the quarters of different tribes or sections of tribes (Prag 1974; Helms 1986:30; Richard and Boraas 1988; Richard: pers. comm). In the Negev, too, several sites have been excavated dating to the EB-MB, consisting of groups of domestic building complexes, such as Har Yeruham (Gophna 1992:134) or Beer Resisim (plan in Finkelstein 1995:93,95). Schloen's (2001: 329-331) analysis of the Late Bronze town plan of Ugarit also reveals a division into insulae of several houses, usually sharing a tomb and a well, and inhabited by what Schloen suggests may have been kinship groups.

Iron Age Stratum III and II of Khirbet Msas, in the Beersheba valley also show a lay-out consisting of clusters or compounds of dwellings, not built in a clearly defined architectural style: some clusters contained mostly four-room buildings (area A; Fritz and Kempinsky 1983: plan 2), whereas other areas contained mainly courtyard buildings. This arrangement certainly seems to indicate the presence of different groups, possibly sections of tribes, that lived together, but in separate sections of the town, and that may have had different building traditions. It may well be related to its proposed function as a way station on the trade route through the Beersheba valley (Finkelstein 1995:67,118, with references; see also Bienkowski and van der Steen 2001, for the role these stations may have played in a tribal society).

Within this patchwork of tribal sections and territories the specific function of the town is that of a marketplace, and the place for redistribution of goods. There is nothing new in that. However, it also has the function of a tribal gathering place, where the different tribal powers in the region had their representative groups, where social relations between the tribes were maintained, and to a certain extent, where tribal power relations were determined.

A more difficult question is whether archaeology can reveal the relationships between the town and the countryside, and, more specifically, between the separate groups in the town, and those of the countryside.

In 19th century AD Kerak the social organization of the town reflected that of the countryside, or hinterland: the different groups that inhabited the separate compounds of the town were largely independent farmers and pastoralists, and at least some of them spent part of the year in their territories outside the town. At the same time, they represented the larger (tribal) units of which they formed a part, and as such their presence in the town may have represented a special function.
However, even though the economy of tribes is marked by flexibility, sometimes tribes specialize: the Rwala were horse breeders, the Howeitat and the Sherarat were traders. The Haddadin in Salt were blacksmiths. Tribes sometimes specialized in certain crafts, for example in Jauf, which was famous for its stone coffee grinders and its specially woven abbayas. Aspects of different specialisms, such as the need for basic materials, or specific ecological niches may have determined both a tribe's territory and its specialism.

The 'pots and people' question needs to be asked again: can certain groups be identified by their material culture, and if so, which aspects of the material culture should we look at? Can artifacts tell us something about the organization of economic interaction between the town and the countryside? Schloen suggests that the inhabitants of Late Bronze Ugarit were generally involved in farming, as shown by the material culture (Schloen 2001:335), a picture that is similar to that of pre-modern Kerak, and that suggests interaction between the town and the countryside for Late Bronze Age Ugarit. It seems likely that this interaction was also organized according to tribal, or kinship-based connections.

Tribal organization is traditionally looked for in burial sites, especially isolated ones. Studies have related the organization of cemeteries to a kin-based organization of the living population that created them. Rast and Schaub state that "In principle it is possible to generate explanations for settlement on the basis of tomb materials" (Rast and Schaub 2003:62). In general multiple burials, whether primary or secondary, are taken as evidence for a family- or clan-organized society (Schaub and Rast 1989:395; Harrison 2001:226; but see Kletter 2002:37 for a cautionary note); but relating this to the architectural organization of the related town remains difficult. The charnel houses in Bab edh-Dhra Str. III, with multiple interments, coincide with the first communal building projects on the site: a temple and a possible tower and town wall. Instead of taking these structures as evidence for a stratified society, with an administrative elite, we may view it as a community made up of different corporate groups, who created communal space (the temple area) as an interaction zone, and a communal defense system. This would fit in well with the fact that the burials show no hierarchical differentiation, while at the same time the charnel houses signify the presence of separate groups. Unfortunately not enough remains of the domestic area to confirm (or refute) this hypothesis, but Harrison suggests that the kin-based organization of Bab edh-Dhra must have continued throughout the 'town-phase' into the EB IV (Harrison 2001:233).

Hallote (2001) has argued for a kin-based society in the Middle Bronze Age, using the Megiddo house burials as as a case-study. In spite of the fact that most of these burials contain only one individual, and none more than three, the fact that they were incorporated in the domestic living space of one family or kin-group, suggests a kin-based social organization of the town (also Schloen 2001:329-31).

Recently Kletter (2002) has asked the question where the IA I people in the highlands buried their dead, since no burial sites of that period have been found in the highlands. Kletter's hypothesis is that there was a change in burial customs after the Late Bronze Age, and the burials of the Iron Age may have been too shallow, and not endowed with burial gifts, and have therefore not been found, or not been recognized. Another possibility is that these people were still buried in their ancestral burial grounds, or caves, which may even have been on the east side of the Jordan. Burials in some of the multiple burial caves east of the Jordan show more of a continuity than the settlement history of the area (Harding and Isserlin 1953, Thompson 1986; McGovern 1986:338; Dajani 1966; Dorneman 1983:31,32).

Finding arguments for a kin-based society in burial customs is not particularly revolutionary. However, relating the population of the settlements, towns or villages, to the population of the countryside on the basis of these burials, requires a more integrated approach to settlement planning, burial sites and material culture than is generally attempted, and may well prove impossible. At the same time, the cemeteries of Bab edh-Dhra or Jericho in the Early Bronze Age, the house burials of Middle Bronze Age Megiddo and Late Bronze Age Ugarit, or the burial caves of the Late Bronze and Iron Ages could reflect a tribal organization that did not stop at the walls of a town, either inside or out.

Bibliography

Abujaber, R.
1989 *Pioneers over Jordan.* London: Tauris.
Bienkowski, P.,and van der Steen, E.J.
2001 Tribes, Trade and Towns: a new framework for the Late Iron Age in Southern Jordan and the Negev. *Bulletin of the American Schools of Oriental Research* 323:21-47.
Burckhardt, J.L.
1822 *Travels in Syria and the Holy Land.* London: J.Murray.
1831 *Notes on the Bedouins and Wahabis, collected during his travels in the East.* London: Colburn & Bentley.
Chesson, M.
2003 Households, houses, neighbourhoods and corporate villages: modeling the Early Bronze Age as a house society. *Journal of Mediterranean Archaeology* 16.1:79-102.
Coogan, M.D.
1984 Numeira 1981. *Bulletin of the American Schools of Oriental Research* 255:75-81.
Dajani, R.W.
1966 Jebel Nuzha tomb at Amman. *Annual of the Department of Antiquities of Jordan* 11:48-52.
Dissard, J.
1905 Les migrations et les vicissitudes de la tribu des 'Amer. *Revue Biblique* 2:410-425.
Dornemann, R.H.
1983 *The archaeology of Transjordan in the Bronze and Iron Ages.* Milwaukee: Milwaukee Public Museum.
Euting, J.
1896-1914. *Tagbuch einer Reise in Inner-Arabien, vol I and II.* Leiden:Brill.
Finkelstein, I.
1995 *Living on the fringe: the archaeology and History of the Negev, Sinai and neighbouring regions in the Bronze and Iron Ages.* Sheffield: Sheffield Academic Press.
Fritz, V., and Kempinski, A.
1983 *Ausgrabungen auf der Khirbet el-Msas (Tell Masos).* Wiesbaden: Harrassowitz.
Gophna, R.
1992 The Intermediate Bronze Age. Pp 126-158 in *The archaeology of Ancient Israel*, ed. *A.* Ben-Tor. New Haven: Yale University.
Greenberg R.
2003 Early Bronze Age Megiddo and Bet Shean: discontinuous settlement in sociopolitical context. *Journal of Mediterranean Archaeology* 16.1:17-32.
Gubser, P.
1973 *Politics and Change in el-Kerak, Jordan.* London: Oxford University.

Hallote, R.S.
2001 Tombs, cult and chronology: a reexamination of the Middle Bronze Age strata of Megiddo. Pp 199-214 in *Studies in the Archaeology of Israel and neighbouring lands in memory of Douglas L. Esse*, ed. S.R. Wolff S.R. Atlanta: American Schools of Oriental Research.

Harding, G.L., and Isserlin, B.S.J.
1953 An early Iron Age tomb from Madaba. *Palestine Exploration Fund Annual* 6:27-47.

Harrison, T.P.
2001 Early Bronze Age social organization as reflected in burial patterns from the southern Levant. Pp 215-236 in in *Studies in the Archaeology of Israel and neighbouring lands in memory of Douglas L. Esse*, ed. S.R. Wolff S.R. Atlanta: American Schools of Oriental Research.

Helms, S.W.
1986 Excavations at Tell Umm Hamad, 1984. *Levant* XVIII,25-50.

Herzog, Z.
1997 *Archaeology of the city. Urban planning in ancient Israel and its social Implications.* Tel Aviv: Tel Aviv University Monograph series 13.

Jaussen, A.
1908 *Coutumes des Arabes au pays de Moab.* Paris: Gabalda.

Joffe, A.H.
1993 *Settlement and Society in the Early Bronze I & II Southern Levant.* Sheffield: Sheffield Academic Press.

Kletter, R.
2002 People without burials? The lack of Iron I burials in the Central Highlands of Palestine. *Israel Exploration Journal* 52:28-48.

LaBianca, Ø.S.
1999 Salient features of Iron Age tribal kingdoms . Pp 19-23 in *Ancient Ammon*, eds. B. MacDonald, and R.W. Younker. Leiden: Brill.

Lewis, N.N.
1987 *Nomads and settlers in Syria and Jordan, 1800-1980.* Cambridge: Cambridge University.

McGovern, P.E., ed.
1986 *The LBA and EIA of Central Transjordan: The Baq'ah Valley Project.* Philadelphia: University Museum.

Mittmann, S.
1994 Hirbet ez-Zeraqon: Eine Stadt der frühen Bronzezeit in Nordjordanien. *Archäologie in Deutschland*:10-15.

Musil A.
1907-8 *Arabia Petraea.* Wien: Alfred Holder.
1927 *Arabia Deserta, a topographical itinerary.* New York: American Geographical Society.

Oppenheim, M.
1943 *Die Beduinen, band II.* Leipzig: Harrassowitz.

Peake, F.G.
1958 *History and tribes of Jordan.* Coral Gables: University of Miami.

Philip G.
2001 The Early Bronze I-III Ages. Pp 163-232 in *The archaeology of Jordan*, eds. B. MacDonald; R. Adams; and P. Bienkowski. Sheffield: Sheffield Academic Press.

Prag K.
1974 The intermediate Early Bronze - Middle Bronze Age: an interpretation of the evidence from Transjordan, Syria and Lebanon. *Levant* 6:69-116.

Rainey, A.F.
2001 Israel in Merenptah's inscription and reliefs. *Israel Exploration Journal* 51:57-75.
2003 Amarna and later: aspects of Social History. Pp 169-187 in *Symbiosis, Symbolism and the power of the past,* eds. W.G. Dever, and S. Gitin. Winona Lake: Eisenbrauns.

Rast, W.E., and Schaub, R.Th.
2003 *Bab edh-Dhra. Excavations at the Town site (1975 - 1981).* Winona Lake: Eisenbrauns.

Richard, S., and Boraas, R.
1988 The Early Bronze IV fortified site of Khirbet Iskander, Jordan: third preliminary report, 1984 season. *Bulletin of the American Schools of Oriental Research supplement* 25:107-130.

Rowton, M.B.
1973a Autonomy and nomadism in Western Asia. *Orientalia* 42,247-258.
1973b Urban autonomy in a nomadic environment. *Journal of Near Eastern Studies* 32:201-15.
1974 Enclosed nomadism. *Journal of the Economic and Social History of the Orient* 17:1-30.

Salzman, Ph.C.
1980 Processes of sedentarization as adaptation and response. Pp. 1-19 in *When Nomads Settle*, ed. Ph. C. Salzman. New York: Praeger.

Schaub, R.T., and Rast, W.E.
1989 *Bab edh-Dhra. Excavations in the cemetery directed by Paul W. Lapp.* Winona Lake: Eisenbrauns.

Schloen, J.D.
2001 *The House of the Father as fact and symbol: Patrimonialism in Ugarit and the ancient Near East.* Winona Lake: Eisenbrauns.

Seetzen, U.J.
1854-59 *Reisen durch Syrien, Palaestina, Phoenizien, die Trans-Jordan länder, Arabia-Petraea und Unter Ägypten.* Berlin: Reimer.

Thompson, H.O.
1986 An Iron Age tomb at Madaba. Pp 331-363 in *The archaeology of Jordan and other studies, presented to S. Horn*, eds. L.T. Geraty, and L.G. Herr. Berrien Springs: Andrews University.

Van der Steen, E.J.
2005 The sanctuaries of Early Bronze Ib Megiddo: evidence of a tribal polity? *American Journal of Archaeology* 109, 1-20.

Wallin, G.A.
1979 *Travels in Arabia (1845 and 1848).* Cambridge: The Oleander Press.

THE ECONOMY OF THE EARLY BRONZE IV PERIOD (*CA.* 2200-2000): THE LITHIC EVIDENCE

Jacob Vardi, Steven A. Rosen and Sorin Hermon

Abstract

The Early Bronze Age IV in the Levant has been the focus of numerous studies. However the flint assemblages have remained largely unknown. A recent analysis of the flint assemblages from Ein Ziq and from Beer Resisim in the Negev Highlands provides some new insights on the basic economic organization of the period in the Negev.

Introduction

The past 50 years of archaeological research in the southern Levant have revealed two distinct periods of florescence in the Negev highlands in the third millennia BCE, the first during the Early Bronze Age II (*ca.* 3000-2700 BCE) and the second during the Early Bronze Age IV (*ca.* 2200-2000 BCE), also referred to as the Middle Bronze Age I (MBI, e.g. Cohen 1999), the Intermediate Bronze Age (IBA, e.g. Amiran 1978), and the Early Bronze-Middle Bronze Age (EB-MB, e.g. Kenyon 1960) (for a summary of the EB IV period terminology, see Gophna 1992: 125-8, 134-6). Given that one of our conclusions is the existence of a basic continuity in the lithic traditions throughout the third millennium BCE, we adopt the term 'Early Bronze Age IV' in this paper.

The origins of the Early Bronze IV are obscure. In the second half of the third millennium BCE the Early Bronze II-III urban centres of the southern Levant were abandoned, replaced by a different settlement system consisting primarily of unfortified rural villages (e.g., Dever 1995; Palumbo 1991: 48-51). Although various reasons for this shift from a more complex society to a simpler level of social organization have been suggested (e.g., Dever 1989), a developing consensus suggests that the change was caused by a gradual climatic warming in the southern Levant (e.g., Neev and Emery 1995:59-65; Miller-Rosen 1995:27-29; Bookman *et al.* 2004).

During the Early Bronze IV period subsistence in southern Canaan was based primarily on pastoralism, with perhaps a variety of domesticated animals including sheep/goat but also cattle and pigs, and cereal agriculture in the northern part of Israel (Dar 1977; Hakker-Orion 1977; Dever 1995: 295; Palumbo 1991, 2001; Cohen 1999: 288-91, 265-6). This economic distinction between north and south was also found by earlier scholars who ascribed it to the difficulties of achieving cereal agriculture in the desert of the central Negev (Aharoni 1978; Glueck 1968:97-101).

In surveys conducted during the past 20 years (especially the Negev Emergency Survey 1979-1982) in the southern Levantine arid regions over 1,000 sites attributed to the Early Bronze Age IV have been documented in the Negev highlands and in the Sinai peninsula (Haiman 1992: 123-4, 1996: 1-5, 23-27). Most of these sites have been described as small encampments containing 2-3 single room structures, usually covering an area of less than 100 square meters (0.1 dunam). Haiman further describes medium (1-2 dunams) and large (2-20 dunams) sites, the latter ones actually small villages (Haiman 1996: 1-5; Cohen 1999: 83-5).

Excavations have been conducted in most of the central sites of the Negev Highlands, including Mashabe-Sade, Beer Resisim, Nahal Nizzana, Beer Hai'l, Har Zayad and Ein Ziq (Cohen 1999: 85, map 4). These sites include, in most cases, several clusters of rounded or oval rooms, each 2-3 m in diameter, attached to one another and thus creating a beehive-like architectural scheme (Cohen 1999: 137, fig. 88; 200, fig. 128). The walls are built of large fieldstones, and often preserved to a height of two to three courses. Each room has a single entry and 1 to 2 stone pillars for roof support (Cohen 1999: 277-9, figs 163-6).

The material culture recovered in these villages contains evidence of a diverse multi-resource economy, typical of societies of the Negev during the third millennium BCE (e.g., Rosen 2003 for the Early Bronze Age I-II). The material culture includes ceramics, milling stones, copper objects, shell beads, and chipped stone tools and waste. The ceramic assemblages show medium size repertoires of vessels with high frequencies of storage ware. Petrographic evidence based on pottery samples taken from some of the sites suggests that the inhabitants maintained trade connections of varying degrees with both close and distant regions, such as the northern Negev, the Judean hills, and central and southern Jordan (Goren 1996: 61-3).

Grinding stones were found in almost every room (Cohen 1999: 266). These reflect both the importance of food processing on the domestic level, as well as a system of production and exchange (see discussion below). A large corpus of metal objects, including weaponry and copper ingots, suggests involvement in a copper exchange system. Haiman (1992:154, 1996:18-20) and Cohen (1999:260-265, 290-291) suggest that the inhabitants of these sites were also engaged in actual metal production.

The Early Bronze IV flint industry from the Negev Highlands has been little investigated. Although the excavations in the central sites yielded a large quantity of flint artefacts, apart from Gilead's (1973) description of the flint implements from Har Yeruham, no reports are available and Cohen's (1999: 268-9) comprehensive study of the period in the Negev includes only a single paragraph describing the lithic assemblages.

In contrast, the past decade has seen both an increase in our knowledge concerning lithic assemblages in the Early Bronze Age I-II in the Negev Highlands, and a greater understanding of their analytic potentials. Thus, information concerning on-site and off-site manufacture, exchange systems, technological systems, functions and economy, chronology, and even cultural attribution has been garnered from various studies. For example, waste and raw material studies have indicated that production was predominantly local and on-site, with only limited evidence for tool exchange or import (e.g. Rosen 1997: 32-34). Concentrations of abraded or exhausted tools within the habitation units suggest domestic use of the tools in such activities as cutting, reaping, butchery (tabular scrapers), and leather work (e.g., Odell 2000 for methods, McConaughy 1979, 1980). Arrowheads indicate hunting or even small-scale warfare and concentrations of microlithic drills suggest semi-specialized industry focused on the production of beads (Rosen 1994-5: 151-5, fig. 2, 2003; Rosen *et al.* n.d.).

Studies about Early Bronze IV flint Assemblages

Flint assemblages from northern Canaan

The Early Bronze IV flint assemblages of the southern Levant in general and from desert regions in particular have been little studied. The main problem with most of the reports about Early Bronze IV lithic assemblages from sites north of the Beer Sheva Basin is that most derive from multi-period sites (tells), for example the materials from Lachish (Waechter 1958) and Jericho (Crowfoot Payne 1983). Assemblages from tells suffer from significant problems of intrusions and determining the specific culture-phase attribution of many non-diagnostic artefacts may be impossible.

Apart from this, flint implements have not always been reported. In other cases, where the flint artefacts have been reported, the focal point has been Canaanean blade technology, in its various manifestations. Other tool types are rarely discussed (Palumbo 1991:112-116, 2001: 257-8).

Given these limitations, the main characteristic of the lithic assemblages from northern Canaan is the continued use of Canaanean blade technology. Directly related to this, sites in the north show a high frequency of sickle blades, both on Canaanean and on simple blades, usually constituting around 20% and sometimes even more of the tool assemblages. The assemblages from the Early Bronze IV site of Sha'ar Ha-Golan (Rosen 1997: 126-9,Table 5.6, Fig. 5.9) and from other sites found in the northern part of Canaan (including Jordan) such as Tell Umm-Hamad (Prag 1974: 97, 99; Betts 1992: 126-7, Table 18), Tell Esh-Shuneh North (Baird 1987, 2001), and Tell Iktanu (McCartney 1996), Nahal Alexander site (Dar 1977), Manahat (Rosen 1998: 79, Table 6.1), Lachish (Waechter 1958: 326), and Jericho (Crowfoot Payne 1983:723) all reflect similar tool assemblage configurations, with ad hoc implements used for a wide variety of domestic tasks and a relatively large number of sickles (glossy blades associated with cutting various kinds of plants and usually associated with reaping [Anderson 1980, Unger-Hamilton 1985]).

Technologically, the northern assemblages are dominated by the production of flakes and flake tools, produced locally with no evident standardization either in terms of production or in the flake tools. Nevertheless, all the reports on northern Early Bronze IV assemblages document relatively high proportions of Canaanean blades and tools – long wide prismatic blades, usually made of fine grained Eocene flint, requiring high skill to manufacture and most often utilized as sickle segments and knives. Canaanean blades and blade tools were clearly the products of some kind of specialized production and distribution system in northern Canaan whose southernmost extent during the Early Bronze IV phase is the site of Jebel Qa'aqir (Dever 1970:147-8, fig. 3) in the southern Hebron Hills. In general, as in the earlier phases of the Early Bronze Age, during the Early Bronze IV Canaanean technology does not appear in the more arid parts of the Negev, the central Negev Highlands and southern Negev.

Flint assemblages from southern Canaan

With the exception of a few general remarks based on preliminary observations of several assemblages from the Negev and Sinai (e.g., Har Yeruham [Gilead 1973], Nahal Nizzana [Rosen n.d.], and for Sinai see Kozloff 1972-3: 40, Rosen 1990), prior to the current research very little was known about the lithic industries of the Early Bronze IV sites in these arid regions (Rosen 1997: 111). Gilead (1973) noted the relative abundance of notches and denticulates and also identified arched backed blades and a single awl found at the site of Har Yeruham. He also noted the relatively high proportion of hammer stones found at the site. In a brief unpublished report on the flint industry from the Early Bronze IV site of Nahal Nizzana excavations that were conducted by Cohen, Rosen (n.d.) described the flint industry as an "ad hoc" industry whose main products are highly amorphous non-standardized flakes, with a limited tool assemblage containing primarily denticulates and miscellaneous trimmed pieces. Notably, neither report indicated the presence of tabular scrapers.

Two other studies of small sites in the Üvda Valley in the southern Negev, with ceramic materials dating both to the Early Bronze I-II and Early Bronze IV, have dealt with the chipped stone materials. The problem of distinguishing the assemblages from the different periods rendered analysis difficult, but spatial analyses of loci which could be separated stratigraphically and architecturally, indicated that macro-lunates which are a subtype of arched backed blades (fig. 7: 4) (e.g. Rosen 1997: 62, fig. 3.21) and tabular scrapers were rare or missing from Early Bronze IV loci (Rosen 1990; 2001: 115-9). Other Early Bronze IV lithic assemblages from sites recorded or excavated during the Negev Emergency Survey (1979-1982) are only briefly mentioned by Cohen (1999: 265-6), without reference to either typological or technological characteristics.

Our knowledge of the lithic industries of Sinai during the second half of the third millennium BCE is even more limited than our knowledge about their counterparts in the Negev. Kozloff (1972-3: 40) claims that they are dominated by crude scrapers, denticulates, and fan scrapers, but acknowledges the problem of intrusions in surface collections. No mention is made of blade tools.

The new data presented here derive from the study of two assemblages from two Early Bronze IV sites located in the central Negev Highlands, the site of Horvath Beer Resisim and Horvath Ein-Ziq (Fig. 1). These assemblages have been analyzed recently as part of a larger synthesis of third millennium BCE lithic assemblages from the Negev Highlands. The fact that these sites are single period occupations with only minimal intrusions offers an opportunity to better understand the specific characteristics of the lithic industries of the terminal stages of the Early Bronze Age sequence in southern Canaan.

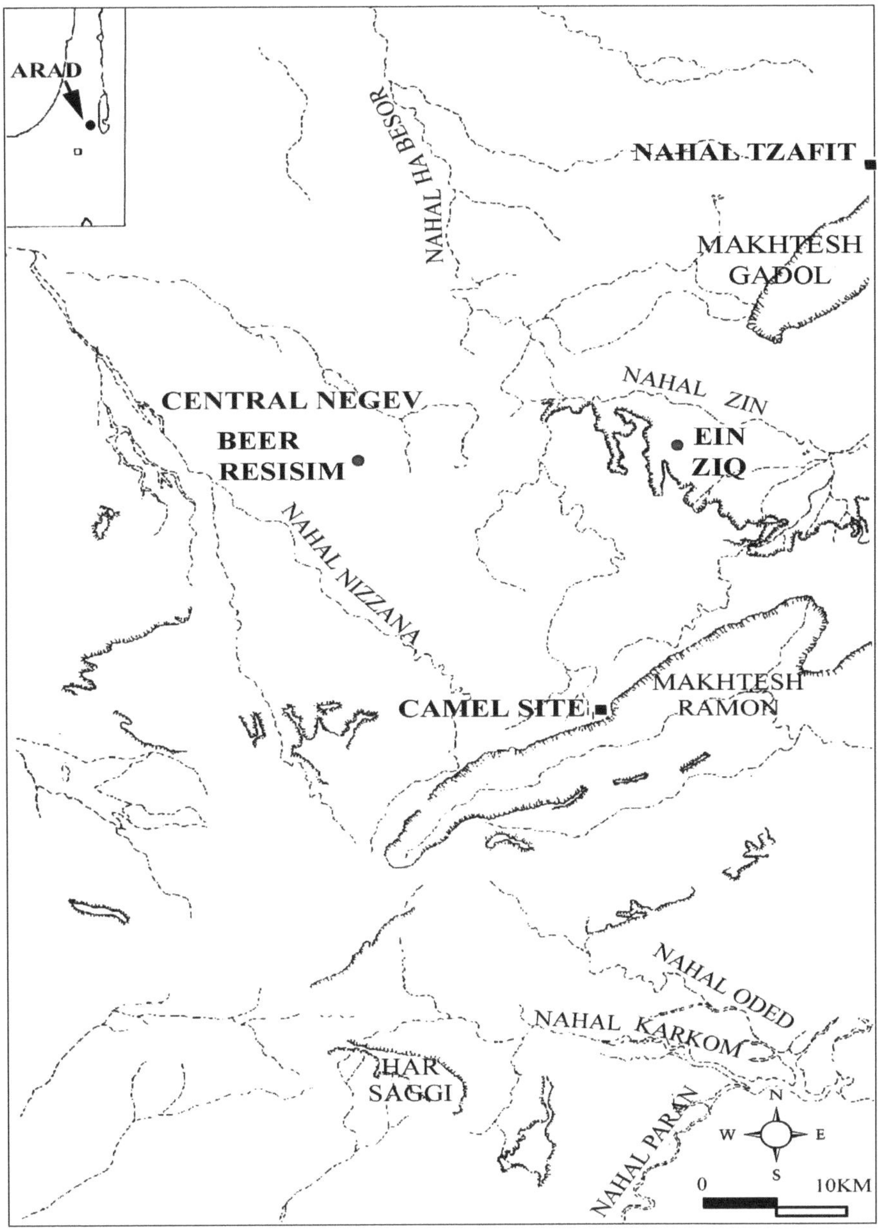

Fig. 1. Sites mentioned in the text: Nahal Tzafit, the Camel Site: EB II, Beer Resisim and Ein Ziq: EB IV.

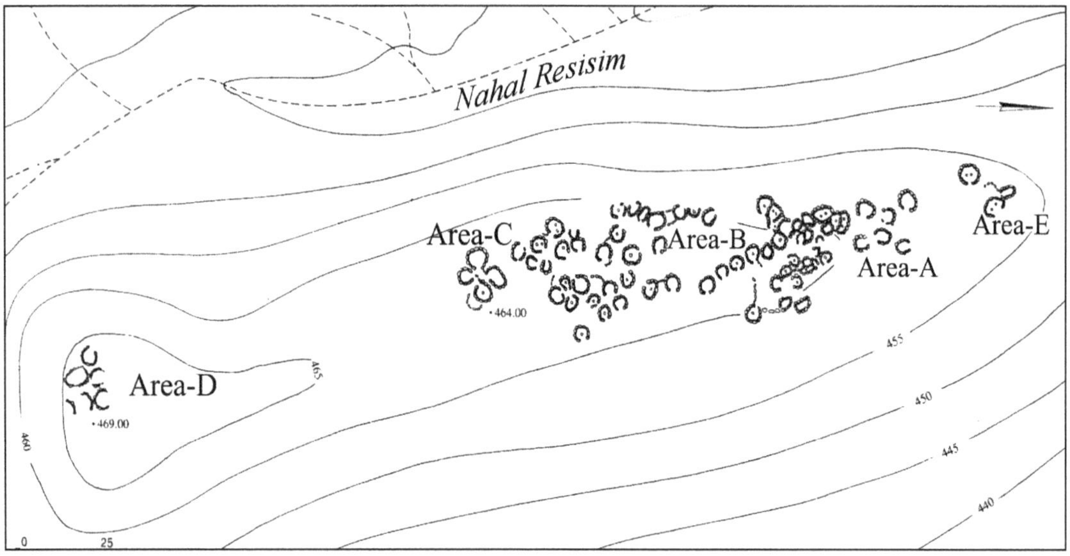

Fig. 2. Beer Resisim site plan 1 (after Cohen 1999: 200, fig. 128)

The Sites

Beer Resisim

Beer Resisim (Fig. 2), approximately 12 dunams in area, is located on a ridge above a ravine called Nahal Resisim in the western Negev Highlands, in an area receiving somewhat less than 100 mm of yearly rainfall today. Excavations were conducted by Rudolph Cohen and William Dever, who led two expeditions between 1978 and 1980 (Cohen and Dever 1978, 1979, 1981a, 1981b). The site consisted of over 100 structures organized in clusters of round or oval units abutting one another, of which approximately 80 percent were uncovered (Cohen and Dever 1981a: 58; Cohen 1999: 201).

Ein Ziq

Ein Ziq is located in the central part of the Negev Highlands, on the northern bank of Nahal Ziq, a tributary of Nahal Zin. There are two perennial springs, Ein Ziq and Ein Shaviv, close to the site, an important element in site location given that rainfall averages only *ca.* 90 mm per year in the area. The excavations of Ein Ziq (Fig. 3), conducted between the years 1984-1986 under the direction of Rudolph Cohen, revealed more then 230 habitation units organized in clusters similar to those of Beer Resisim (Cohen 1999: 137-8). The site extends over an area of approximately 20 dunams, making it the largest known single period Early Bronze IV site in the Levant.

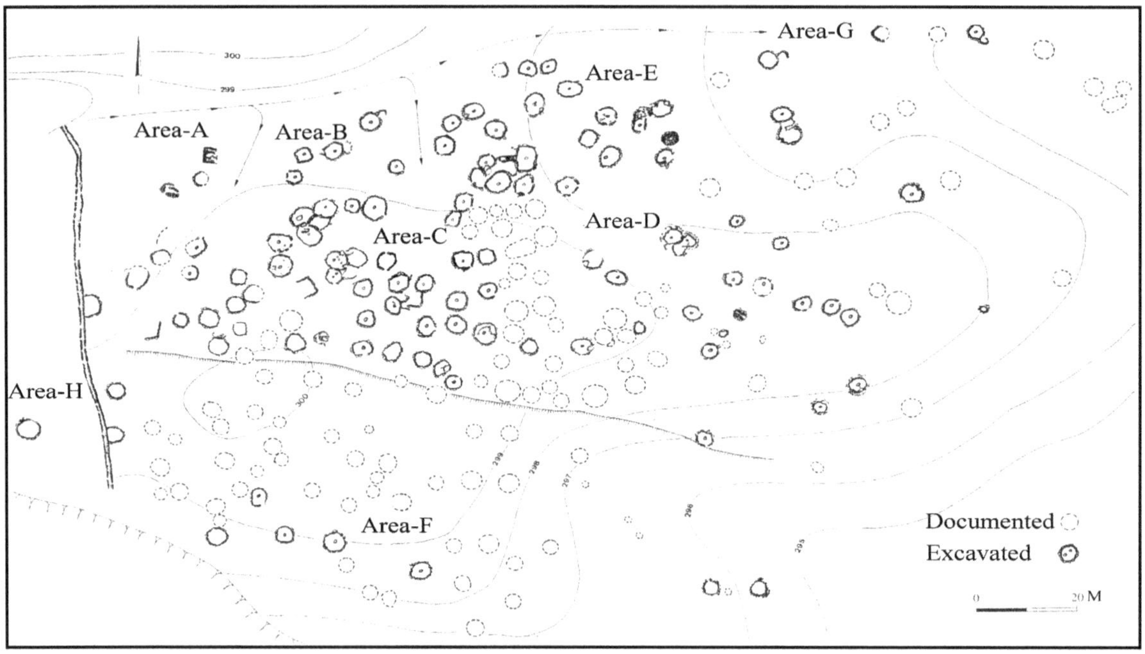

Fig. 3. Ein Ziq site plan (after Cohen 1999:137, fig. 88)

The excavation and collection methods
Excavations were conducted at both sites using the locus-basket method, where the structures themselves served as the basic provenience units. Thus room contours were identified visually and excavations were conducted within the inner part of the rooms (Dever p.c.) while the "alleys" or passages between rooms remained untouched. This constitutes a significant drawback in analyzing lithic discard patterns since large parts of the sites were, in fact, not excavated.

Indeed, in most cases each room received a locus number and a single basket number referring to all the sediments from the site surface down to the occupation surface. No distinctions were made either stratigraphically or areally within the rooms. Artefacts were collected manually and sieving was conducted only in special cases. There was no sieving of excavation sediments (Dever *pers. comm.*; Israel *pers. comm.*). The failure to sieve consistently, and the consequent absence of the smaller range of artefacts, affects the frequency of some waste and tool types. However, based on the major waste groups and the tool group proportions compared to the total assemblage, the assemblages are fit for a thorough analysis (Rosen 1997: 37-8).

The Lithic Assemblages
The lithic assemblages were divided according to the tool and waste typologies standard in lithic analysis or prehistoric and proto-historic assemblages in the Levant (e.g., Tixier 1963; Bar-Yosef 1970; Gilead *et al.* 1995; Rosen 1997). The use of these standardized typologies provides a clear description of the assemblages that allows a base for future comparisons with other assemblages and industries from both the Negev and other regions (Gilead *et al.* 1995; Rosen 1997).

The Lithic Assemblage from Beer Resisim

The lithic assemblage from Beer Resisim consists of 12,098 artefacts, of which 931 (7.7%) are tools (Table 1).

Raw material
The raw material from Beer Resisim originates primarily in the large flint nodules that are found loose or partially embedded in the chalky bedrock of the site. Some of these flint nodules are fairly large (~70 cm max. dimension). The colour of the raw material used varies from light to dark grey and from light to dark brown. The texture of the raw material varies also from fine to medium grained flint. In some cases, this range of colours and textures appears in the same artefact. It is intriguing to note that the nodules described are suitable for the production of tabular scrapers, but that in spite of this, the frequency of tabular scrapers is minimal, and only six are typical tabular scrapers (see discussion below).

The waste
Flakes: (n=4715, 39.0% of the assemblage): the flakes are the most dominant product found in the assemblage. A sample of 100 flakes has been tested for various technological observations and metrical analysis (Table 2). The flakes from Beer Resisim are irregular in shape. Most of the flakes have either flat (n=65) or cortical (n=23) butts. Only a few of the flakes tested show some level of core modification: four are facetted, two have a dihedral butt and one is punctiform (Inizan *et al.* 1992: 79-81, fig. 32), while the remaining five butts are broken or crushed (indeterminate). More than a quarter of the sampled flakes (n=28) end with a hinge fracture, probably indicating a low level of expertise (Inizan *et al.* 1999: 36-37; Whittaker 1999: 17-18).

	Ein Ziq			Beer Resisim		
Debris	N.	%-of Group	%- of Total Assemblage	N.	%- of Group	%- of Total Assemblage
Chips	1,633	47.7	13.7	3,239	65.9	26.8
Chunks	1,788	52.3	15.0	1,678	34.1	13.9
Total Debris	*3,421*	*100.0*	*28.8*	*4,917*	*100.0*	*40.6*
Debitage	N.	%-of Group	%- of Total Assemblage	N.	%- of Group	%- of Total Assemblage
Flakes	5,962	81.2	50.1	4,715	78.3	39.0
P.E	1,075	14.6	9.0	900	15.0	7.4
Blades& Bladelets	107	1.5	0.9	315	5.2	2.6
C.T.E	196	2.7	1.6	88	1.5	0.7
Total Debitage	*7,341*	*100.0*	*61.7*	*6,018*	*100.0*	*49.7*
Cores	195		1.6	129		1.1
Hammers	72		0.6	103		0.8
Tools	867		7.3	931		7.7
Total Assemblage	*11,895*		*100.0*	*12,098*		*100.0*

Table 1: Waste and tool frequency from Ein Ziq and Beer Resisim

Measurements in mm.	Average	S.D	Min	Max
Length	42.2	14.4	18	101
Width	40.1	13.3	19	86
Thickness	10.2	3.7	4	23
(Length+Width)/2	41.2	12.3	21	88

Table 2: Beer Resisim flake dimensions

Ein Ziq			Beer Resisim		
Type	**N.**	**%**	**Type**	**N.**	**%**
Flake	160	82.1	Flake	84	65.1
Blade	3	1.5	Blade	3	2.3
Bladelet	0	0.0	Bladelet	2	1.5
Mixed:Flake/Blade	32	16.4	Mixed:Flake/blade	40	31.0
Total	195	100.0	Total	129	100.0

Table 3: the frequency core types from Ein Ziq and Beer Resisim

The flake vs. primary element ratio is 5.2:1, which probably indicates a very short sequence of flake reductions – post core decortication. The very few core trimming elements found also attest to a very minimal modification of the cores, if at all. The core trimming elements found are irregular and do not resemble the standardized core tablet and crested blades that appear in earlier periods in more standardized industries.

Blades and Bladelets (n=315, 2.6% of the assemblage): the blade and bladelets from Beer Resisim are quite crude and in some cases they look robust. They are made on the same kind of local raw material as the flakes. An analysis of the blanks width dimensions, in this category, shows that there is no distinct population of bladelets. Most of the blades (n=290, 92%) have irregular scar patterns which means that it is possible that a reduction sequence of flakes was conducted prior to their removal (Gilead *et al.* 1995: 235).

The ratio between the flakes and the blades/bladelets is lower (about 8:1) than at Ein Ziq and earlier Early Bronze I-II sites, reflecting the presence of a relatively high number of blades. This is also apparent in the tool assemblage, which shows a larger number of blade tools of various kinds.

Cores (n=129, 1.1% of the assemblage): the cores from Beer Resisim are in most cases flake cores although compared to Ein Ziq there is a relatively high amount of mixed flake and blade cores (see Table 3) indicating that the manufacturing of elongated blanks had some importance at Beer Resisim. It seems that there are only few indications for core preparation / rejuvenation since the ratio C.T.E: cores is very low: 0.7:1. This is also evident in the amount of cortex found on most cores including on their striking platforms (Fig. 4). The maximum core dimensions (in mm) from Beer Resisim are: mean 67.1±26.2; min 28mm; max 170mm.

Fig. 4. a refitted core from Beer Resisim structure 6, room 5A

The tools

The tool assemblage from Beer Resisim resembles that of Ein Ziq and other EB flint assemblages, in the general dominance of the *ad hoc* tool types. However, it differs in several particulars, most notably the higher number of blade tools, the presence of a cache of microlithic drills, and the presence of a few tabular scrapers.

Borers (n=136, 14.6%): this category includes all tools that exhibit one or more modified points. Most borers (n=125) are awls exhibiting a short bit modified by two adjacent notches (Fig. 7:5) .The remaining nine are drills with elongated bits, made on thick flakes or blades (Fig 7:6). The drills were made by bilateral abrupt retouch, converging into a point (Gilead *et al.* 1995: 245, fig. 5.16:2-4, 7-6; Rosen 1997: 68-70, fig 3.26-3.27). The group labelled as borers is highly non-standardized. The awls and the drills were modified on flakes in most cases, but they also appear on other types of waste. Another proof for the non-standardization of this tool group can be found in the wide size range, ranging from very small awls to heavy-duty awls. The heavy-duty tools seem to form an important size group within the category borers (n=24, 19% of the awls group).

Micro-drills (n=41, 4.4%, Fig. 5): most of the micro drills were found in a single architectural unit (room 22, Area B), where 36 microlithic drills were found in association with a group of shells (Fig. 6). This small bead production atelier is reminiscent of a similar atelier found at the Early Bronze II Camel Site (Rosen 1994-5). The microlithic drills are in fact small drills that appear to be rather homogenous in size and bits dimensions (Fig. 5).

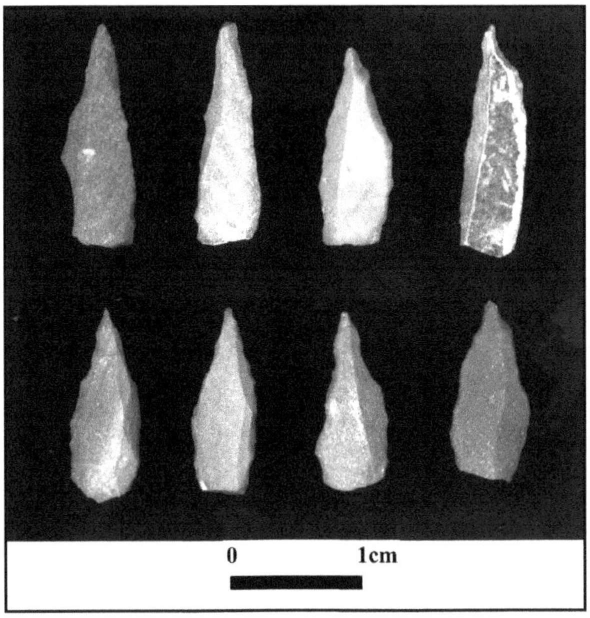

Fig. 5. the micro drills from Beer Resisim

The micro-drills that we tested had an average bit dimension ([width+thickness] /2) smaller than 5mm.This dimensional uniformity is probably the outcome of the function of these micro-drills as drilling instruments in the bead industry, creating a necessity for specific dimensions. All micro-drills (n=40) except two are on narrow blades or bladelet blanks. The two exceptions are made on a small flake and a primary element (Fig. 5). The bits of the micro-drills are elongated and modified by abrupt retouch, with two notches separating the bit and creating a "winged" shaped micro-drill. Most of the micro-drills have their butts modified (basal modification). This probably indicates their usage as bow drills (Anderson 1991, Hughes-Brock 1998, Stocks 1989). It is important to note that the micro-drills from Area B are the only flint artefacts collected from this area, probably due to the special context in which they were found (Cohen and Dever 1981b: 572).

Fig. 6. a concentration of shells found at the provenance of the micro drills from Beer Resisim: Room 22 in structure 11 in area B. (After Cohen 1999:211, fig. 176)

Notches and Denticulates (n=363, 38.9%): this *ad hoc* tool group includes all tools that have one or more concavities (Rosen 1997:90-1). There are 206 (56% of this tool group) single or double-notched tools. 155 tools (43% of this tool group) have 3 to 8 notches, and the remaining two have nine and ten notches respectively, often creating a denticulated working edge (Fig. 7:8). The tools of this group are made primarily on flakes (n=264, 72% of the tool group), the rest are made on other waste types.

Scrapers (n=81, 8.7%): this category consists of scrapers with a scalar retouch which has created a semi abrupt to abrupt active edge (Fig 7:9). It can be described as morphologically heterogeneous, with little standardization in manufacture. The category can be divided into three general subtypes, based on the active edge location (Gilead *et al.* 1995: 237-245 fig 5.11-13; Rosen 1997:86-87, fig. 3.44-5). However there is no indication that such a division has any functional meaning.

1. End scrapers (n=47). With this subtype the working edge covers the distal end (fig. 7:9). Two end scrapers are minute (the largest one measures 39x36x22mm) and two can be defined as heavy-duty end scrapers (the largest one measuring 125x103x33mm).

Fig. 7. Flint tools from Beer Resisim: 1, 4 - Arched back blades; 2, 3 – sickles; 5 – awl; 6- drill; 7- retouched flake; 8 – denticulate; 9 – scraper; 10 – tabular scraper.

2. Side scrapers (n=22). 16 side scrapers have scraper retouch on one side, five are double side scrapers, and one is a minute side scraper (measurements 47x40x23mm).
3. Indetermina (n=12). In addition the scraper category includes also three butt scrapers and nine broken scraper fragments.

Tabular Scrapers (n=6, 0.6%): although the number of retouched flakes that were made actually on a tabular primary element is higher, only six pieces are really tabular scrapers with obvious scalar retouch. One of the tabular scrapers (Fig. 7:10) has the classical fan scraper attributes (Neuville 1934, Rosen 1997:73, fig. 3.30), a modified butt and a wide arched working edge (measurements 93x106x16mm), a second is a tabular side scraper with inverse retouch covering both lateral sides including the butt (70x95x17mm). The third one is leaf shaped, the retouch covering almost the whole tool circumference, and it has butt modification and bulb thinning on its ventral side (97x57x10mm). The fourth tabular scraper also has a leaf shape. Its left side has fine retouch while its right side is bifacially retouched (115x68x17mm). The remaining two tabular scrapers are broken, and specific types could not be determined.

Retouched Blades (n=90, 9.7%): this category includes all blades that have continued retouch covering at least 2 cm. The tools in this category include several subtypes, differentiated on the basis of basic morphological features.

1. Simple retouched blades (n=56). The simple retouched blades are irregular when it comes to their dorsal scar patterns (Rosen 1997: 63, fig. 3.22 8-11). Five simple blades can be categorized as heavy-duty tools: they are quite large (the largest one is 156x56x24mm) and have nibbling on one of their sides. 13 simple blades are truncated. Two of these are bi-truncated (proximally and distally), which suggests possible hafting.
2. Backed blades (n=34). All backed blades have partial or complete backing on one of their sides (Figs 7:1, 7:4). The backing consists of semi-abrupt to abrupt retouch and in most cases is made by direct retouch. The existence of backing probably hints that those tools were hafted (c.f. Rosen 1997: 62-64, fig. 3.22).

Sickles (n=22, 2.4%): this category includes all artefacts that show sickle gloss, which is the outcome of cutting grasses (Anderson 1980; Unger-Hamilton 1985; Rosen 1997:55-7). Morphologically the sickles appear on simple retouched blades (n=4, Fig. 7:3) or on backed blades (n=18, Fig. 7:2,), those were probably hafted (Rosen 1997: 57, 64). Of the backed blades five have a straight, and 13 an arched back. In some cases (n=6) the sickles are quite long: ≥7cm in length, indicating that they were hafted individually, thus the term reaping knifes would, in fact, describe them better (Rosen 1997: 57).

Retouched flakes (n=171, 18.4%): the retouched flakes group is a catch-all category containing all the retouched tools that do not fall into any of the other categories (Rosen 1997: 92-3). Most of the retouched flakes have an irregular retouched lateral or bilateral working edge (fig. 7:7).

Varia (n=21, 2.3%): this category consists of several subtypes such as

1. Burins (n=14). Eight burins have a single burin blow, five of which are found on proximal sides. One of these also has a short point modified by two notches, so it may be a double-function tool (burin/awl). Another burin is on natural pan (Bar-Yosef 1970: 208, fig 5:24). One burin is massive and made on a retouched flake (65x70x31mm), while two other burins are on flake fragments. Another five burins have two burin blow scars. Two are dihedral burins; one is a dihedral angled burin (Bar-Yosef 1970:208, fig 5:23). In two other cases the two burin blows trim the distal or proximal edges. It is possible that the burins from Beer Resisim were not utilized (c.f. Rosen 1997:100)
2. Intrusions (n=3). One backed bladelet obliquely snapped at its distal edge is probably an Epipalaeolithic intrusion. A second intrusive tool is a broken arrowhead also on burnt flint, probably Neolithic in origin. The third is a micro burin, also probably an Epipalaeolithic intrusion.
3. Pièce Esquillée (n=2). Both pieces have inverse reductions on their proximal edges, which may reflect a knapping procedure that involved the use of an anvil (Bar-Yosef 1970:220). It is possible that these were not actually utilized.
4. Marbles (n=2). Two small stone marbles, probably collectibles. Their use is unknown.

The Lithic Assemblage from Ein Ziq

The lithic assemblage from the site of Ein Ziq consists of 11,895 pieces of which 867 (7.3%) are tools (Table 1).

Raw material

The raw material that was used for the production of flint implements is a fine-grained dark brown Eocene flint. It appears in two forms, as amorphous nodules or oval lenses taken from the Ein Mor formation in the local limestone layers (Idelman *et al.* 2000: 112-113). Numerous nodules are found within the immediate premises of the site and in the adjacent ravine. Thus, all the raw materials used for lithic production are local.

The waste

Flakes (n=5962, 81.2% of the debitage, 50.1% of the assemblage) are the most dominant blank produced at Ein Ziq. The flakes are medium in size and appear to be irregular in shape. In most cases, it appears that the flakes are as wide as they are long (table 4). Of a sample of 100 flakes, 25 end with hinge fractures, probably indicating an industry of minimal expertise, which is in accordance with the analysis of the cores. Flake butts are flat in most

Measurements in mm	Average	S.D	Min	Max
Length	39.5	12.0	20	75
Width	38.5	11.8	17	73
Thickness	10.4	3.8	3	23
(Length +Width)/2	39.0	10.1	19.5	65

Table 4: Ein Ziq flake dimensions (n=100)

	Beer Resisim		Ein Ziq	
Major Tool Types	N.	%	N.	%
Borers	136	14.6	167	19.3
Micro-Drills	41	4.4	4	0.5
Notches and Denticulates	363	38.9	324	37.4
Scrapers	81	8.7	37	4.3
Tabular Scrapers	6	0.8	0	0.0
Retouched Blades	90	9.7	36	4.2
Sickles	22	2.4	5	0.6
Retouched Flakes and Pieces	171	18.4	274	31.6
Varia	21	2.3	20	2.3
Total	931	100.0	867	100.0

Table 5: Major tool types frequencies from Beer Resisim and Ein Ziq

cases (n=66); the remaining are cortical (n=9), dihedral (n=6), facetted (n=4), and crushed or broken (n=12) (Inizan *et al.* 1992: 79-81, fig. 32). These frequencies show that almost no modification was carried out prior to the flakes' removal.

The ratio of primary elements (P.E.) to flakes is low (*ca.* 1:6), indicating rapid discard after the initial flake removals. The number of core trimming elements (C.T.E) is very low. Most are non-diagnostic and do not resemble any of the standard core trimming elements, such as ridge blades or core tablets, known from earlier industries. The very low ratio of C.T.E.: cores indicates that core modification before and during the blank production was very limited, in accordance with the low primary element: flake ratio.

<u>Blades and Bladelets (n=107, 1.5% of the debitage, 0.9% of the assemblage):</u> the blades and the bladelets are grouped together, first of all because the number of blades and bladelets is minimal, and second because, based on Tixier's method of distinction between blades and bladelets (12 mm width limit), we did not find any distinct population of bladelets (blades whose width is less than 12mm). Therefore it is likely that the so-called bladelets from Ein Ziq are, in fact, blades. This whole category blanks is irregular in shape. In most cases their dorsal scar pattern is irregular which means that the blank reduced prior to its removal did not have to be a blade or a bladelet (Gilead *et al.* 1995: 235). Thus there is no discrete procedure for the production of blades and bladelets.

<u>Cores (n=195, 1.6% of the assemblage):</u> the cores are mostly amorphous in shape and are usually designated for the production of flakes (Table 3). Some of the cores bear scars both of flakes and blades or bladelets, but only on rare occasions does the number of blade scars exceed the number of flake scars. Each core has between 1-3 striking platforms, which probably reflect the extent of their usage. The cores show very little modification, striking platforms being usually plain, which means that they were decorticated by a single blow removal (cf. Inizan *et al.* 1992: 45-6).

It is worth noting here that partial refitting was conducted of the assemblage of several rooms on both sites, indicating that at least some lithic reduction was carried out within the structures. The refitting supports the "minimal expertise notion". Refitted flake butts are either flat or decorticated, correlating to the plain/cortical striking platforms frequencies.

The tools

The tool assemblage from Ein Ziq can roughly be described as *ad hoc* or expedient in nature. The major tool types that dominate Ein Ziq assemblage are borers, notches and denticulates, as well as retouched flakes. Other tool types such as retouched blades, sickles, and microlithic drills appear in very low frequencies (Table 5).

Fig. 8. flint tools from Ein Ziq: 1- micro-drill; 21 drill; 3, 4- awls; 5- sickle; 6- denticulate; 7- retouched flake; 8, 9- scrapers

Borers (n=167, 19.6%): this group includes drills (n=5, Fig. 8:2) and awls (n=162, Fig. 8:3-4). The drills have an elongated bit which is at least one third of the total tool length. Although they have the same proportions as blades (length ≥ width x 2), it is possible that these proportions are the outcome of the point modification, meaning that they are made on flakes. The awls have a much shorter bit, modified by one or two notches, made on the blank's distal part. They are made in most cases on flakes (but appear also on primary elements and in rare cases even on pebble chunks (n=3) or cores (n=2). This group is highly non-standardized, in terms of dimensions, mode of preparation etc. (Gilead *et al.* 1995: 245, fig. 5.16:2-4, 7-6; Rosen 1997: 68-70, fig. 3.26-3.27).

Micro-Drills (n=4, 0.5%): only four micro drills were found in Ein Ziq. These are small versions of drills (Fig. 8:1). The distinction between the micro-drills and borers (of which they are practically speaking a subgroup) is based on their connection to the bead industry (Burian and Friedman 1985, 1987; Rosen 1994-5, 2003). One of the drills has two wings separated from the bit by two adjacent deep concavities (Fig. 8:1). Another one is made on a semi-translucent flint bladelet whose point modification is by fine semi-abrupt retouch. Although only two of them have their butts modified (retouched opposite to their bit), it is most likely that they were used as bow drills (Andrews 1990: 68-81, Hughes-Brock 1998: 262-264, Stocks 1989 for experimental reconstructions).

Notches and Denticulates (n=324, 37.8%): this group includes tools that have a working edge composed of one or more concavities (Rosen 1997:90-1). Most of the tools of this tool group are notches, with one to two notches on each tool (n=215, 66%). The remainders are denticulates (Fig. 8:6) whose working edge includes 3 to 6 notches, while one artefact has 7 notches. This tool group is highly non-standardized. The notches and the denticulates appear on all types of blanks but they were made on flakes (n=255, 78.7%).

Scrapers (n=37, 4.3%): Ein Ziq has produced very few scrapers (Fig. 8:8-9). All have semi-abrupt to abrupt scraper retouch, which creates a working edge on one or two extremities (Gilead *et al.* 1995: 237-245 fig 5.11-13; Rosen 1997: 86-87, fig. 3.44-5). This group is highly non-standardized: 23 of the scrapers (62%) are made on flakes, the rest appear on primary elements (n=7, 18.2%), core trimming elements (n=3, 8.1%), core (n=1, 2.7%) and pebble chunks (n=3, 8.1%). Although it is customary among prehistorians to divide the scrapers according to their delineation as well, we find it pointless, since there is no evidence for any correlation between this kind of attribute and the tool's function.

Retouched blades and bladelets (n=36, 4.1%): these include all blade tools that have continued retouch of more than 2 cm on their side or distal end. Most of the blades are simple and non-prismatic. They can be divided into several subtypes:

1. Simple Retouched Blades (n=24): this subgroup includes blades with a continued retouch on one or two sides (Rosen 1997: 63 fig. 3.22:8-11). The tool's working edge is only slightly nibbled; the retouch appears either on the dorsal side (direct), ventral side (inverse) or other combinations (alternating, Inizan *et al.* 1992: 94, fig. 44).
2. Backed blades (n=8). eight of the retouched blades are backed. The backing suggests that those tools were hafted (Rosen 1997: 64). On seven of these the back is made by abrupt retouch. Three have a complete arched back, (c.f. Rosen 1997: 62 fig. 3.21:5-9) while five others have a partially retouched back (the backing was always made by direct retouch). One artefact has an almost fully cortical back and partial modification. Three of the backed blades are truncated, convex (n=1), oblique (n=2) and straight. One of the arched backed blades is bi-truncated.
3. Retouched bladelets (n=4): this subgroup includes four bladelets with a width of less than 9 mm (average measurements are 15.25x7.25x2.62mm). Three are truncated and have slight nibbling on one side (this could well be edge damage and not actually retouch). One is a bi-truncated bladelet with no retouch on the sides; this artefact is made on a semi translucent grey flint.

Sickles (n=5, 0.6%): five tools are defined as sickles, all have sickle gloss. In four cases the gloss covers the entire working edge. One sickle was rejuvenated, meaning that "new" retouch was made on the working edge "cutting" the old gloss. Two of the sickles can be morphologically defined as arched backed blades (Fig 8:5). Three others are on simple retouched blades. One of them is actually a reaping knife. It is larger than the rest of the sickles (88x21.5x5mm) and was probably hafted individually (Rosen 1997: 57).

Retouched flakes and pieces (n=274, 31,6%): this is a catch-all category that includes various trimmed pieces, most of which are flakes (fig. 8:7). But there are also retouched primary elements, and several tools that were made on core trimming elements.

Varia (n=20): the varia category consists of several artefacts that either could not be classified in any of the categories, or appear in very small numbers.

1. Burins (n=3): all burins are on flakes, two are on a natural pan (Bar-Yosef 1970: 208, fig. 5:24). It is possible that the burins from Ein Ziq were not utilized (c.f. Rosen 1997: 100)
2. Modified pebbles (n=4): medium fist-size pebbles with parallel bifacial reductions that create a low-angle, sharp working edge.
3. Choppers (n=8): all choppers are on robust pebble chunks. Morphologically they are similar to the modified pebbles, but their working edge bears some bashing marks (Leaky 1971: 4-5, fig. 39 3-5; Rosen 1997: 98-100, fig. 3.52).

4. Pièce esquillée (n=2): both artefacts classified as pièce esquillée are on flakes. They have opposite reductions on their ventral side on the distal edge, which could be the result of a knapping procedure involving the use of an anvil (Bar-Yosef 1970: 220).
5. Working edge rejuvenation (n=2): two pieces have been trimmed of the working edges of tools, by a burin-like side blow.
6. Hand axe (n=1): this lower Palaeolithic tool was probably a collectible and brought to the site from the Avdat plateau to the south of Ein Ziq.

Discussion

Field Methodological Considerations

Collection techniques can significantly affect the configuration of lithic assemblages. The lack of systematic sieving of sediments at Beer Resisim and Ein Ziq results in clear biases towards the recovery of large artefacts. Furthermore, although it appears that such small types as micro lunates and other small arrowheads, along with other retouched bladelet tools, dropped out of the lithic repertoire in the Negev by the middle of the 3rd millennium BCE (Fig. 10), the absence of systematic sieving precludes definitive conclusions. Similarly, the much lower proportion of chips in these assemblages than, for example, in the Camel and Nahal Tzafit sites (Fig. 9), is also probably a reflection of the lack of sieving. This bias towards larger pieces also renders metric comparisons difficult, for obvious reasons.

Typological Configuration

Analysis of the waste and tool groups reflects great structural similarity between the two Early Bronze IV sites examined here and two Early Bronze I-II sites, Nahal Tzafit and the Camel Site (Figs 9-10). Although we are aware that typological analyses cannot automatically be assumed to reflect function, in this case we suggest that the similarities between the frequencies of the major tool categories probably reflect similar functions of the chipped stone tools in these periods.

Any typological description of material culture in the Early Bronze Age sequence has to consider its location within the chronological continuum. The Early Bronze Age IV is a relatively short period within this continuum and it appears, judging from a series of C14 dates from known Early Bronze IV sites that they are closer in time to the Early Bronze II-III than has been assumed until now (Avner *et al.* 1994; Segal 1999: 336-9).

Chronological Implications

The presence of tabular scrapers at Beer Resisim suggests that this tool type either continued into the Early Bronze IV, or that it is intrusive. The very low numbers of tabular scrapers at the site of Beer Resisim and their absence at Ein Ziq (as well as from Har-Yeruham [Gilead 1973] and other sites), together with the general absence of the type at most Early Bronze IV sites further north (Waechter 1958), strongly suggests that the pieces found at Beer Resisim reflect either an earlier and ephemeral Early Bronze Age I-II presence on the site, or collection of

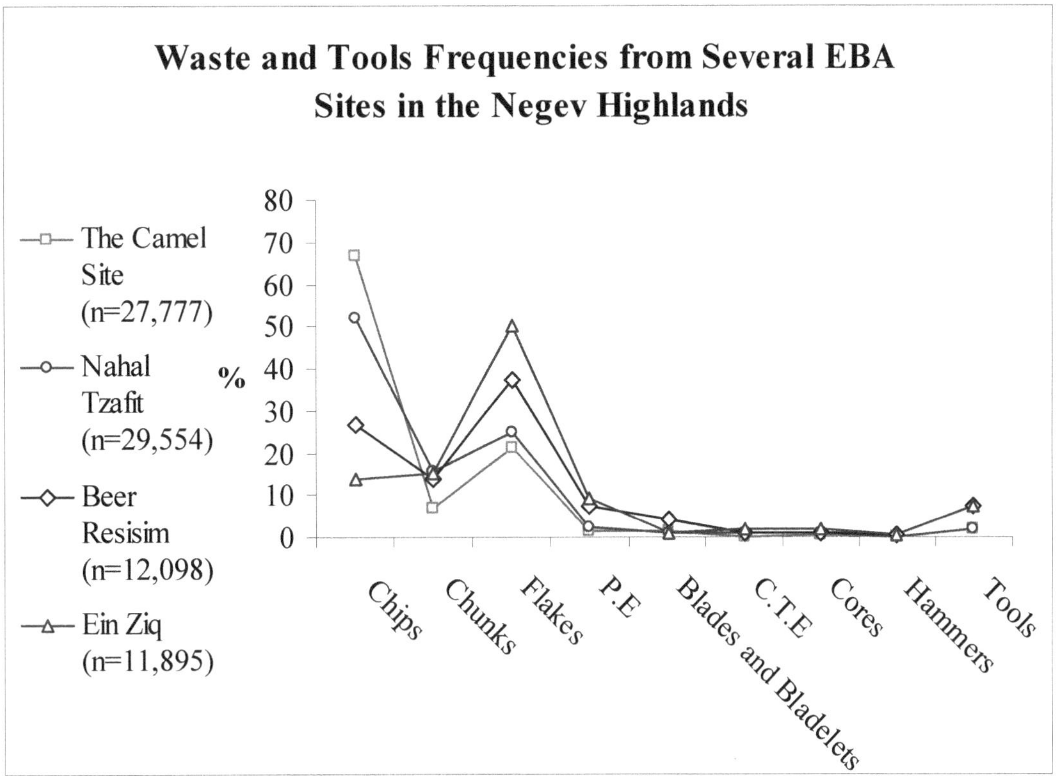

Fig. 9. A comparative frequency of the waste and tools from sites mentioned in the text

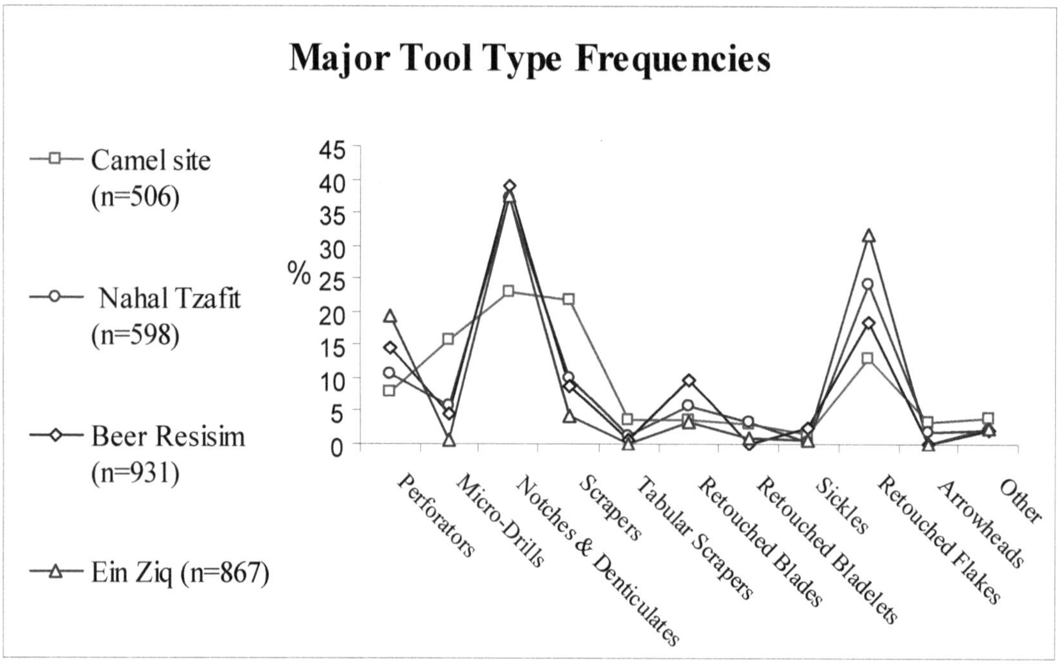

Fig. 10. a comparative tool type frequency graph

these tools from other nearby Early Bronze Age I-II sites. It is important to note at this point that Cohen and Dever's original assumption, based on ceramics found during the three seasons of excavations at Beer Resisim, was that there was solid proof for the existence of Early Bronze II occupation within the site's premises (Cohen and Dever 1979: 57-8). This view was changed later by Cohen (1999: 201), who claimed that similar pottery vessels were found in later excavations conducted at other places.

In contrast, the presence of arched backed blades, manufactured on local flint, as shown by the associated production waste, suggests strongly that these can be associated with the Early Bronze IV occupation, contrary to earlier assessments attributing them to the Early Bronze Age I-II.

Economy

According to use ware analysis checks run by McConaughy on flint implements from the Early Bronze Age site of Bab edh-Dhra (McConaughy 1979, 1980) it appears that the main bulk of tools that appear in the assemblages of Ein Ziq and Beer Resisim were used for a wide variety of domestic tasks. Those activities included the processing of organic materials such as plants, wool (shearing), and leather preparation. Some of the tools were utilized for the modification of harder materials such as wood, bone and stone etc. The presence of a large quantity of flint products from all stages of the chain of operations is significant. Both sites provide solid proof for the existence of a dynamic chipped stone industry on their premises. In other words: the flint industry at Ein Ziq and Beer Resisim was domestic and does not contain any indication of imported artefacts.

Technologically one fact distinguishes the northern Early Bronze IV assemblages from their southern cousins – the total absence of Canaanean blade technology for the manufacturing of retouched blades, reaping knifes, and sickle segments (compare for the north e.g., Dar 1977; McCartney 1996; Rosen 1982, 1983). As indicated above, the southernmost site in which Canaanean blades were found in Early Bronze IV contexts is the site of Jebel Qa'aqir which is located at Southern Hebron mountains (Dever 1970).

This dichotomy also obtains for the Early Bronze I-II, when the distribution of Canaanean blades was somewhat further south, as far as Arad, but beyond which blade technologies were dominated by less sophisticated production techniques. The difference in distribution reflects the difference between the settlement systems of the different periods, rather than some inherent difference in the distribution of the Canaanean blades themselves (Rosen 1997: 58-9, fig. 3.18). Thus, the Canaanean blade production and exchange system reflects basic continuity throughout the third millennium BCE, including continuity in the limits of its distribution. In this context, it is intriguing to note that petrographic analyses of ceramics from Negev Highlands Early Bronze IV sites, including Ein Ziq and Beer Resisim, indicate exchange relations with northern regions such as the Judean hills and the Shephela, as well with Wadi Zarqa, on the central Jordanian plateau (Goren 1996). Thus, the absence of Canaanean technology contrasts with the presence of other aspects of material culture deriving from northern areas.

The presence of bitumen on the back of one sickle from Beer Resisim (and in other Early Bronze IV sites in general), originating in the Dead Sea and within the sphere of Canaanean technology, accentuates the contrast even more. Bitumen was a common adhesive used for Canaanean blades in both the Early Bronze IV and earlier periods (e.g. Schick 1978, McCartney 1996), and for flint sickle segments in general in ancient times, and itself generated exchange systems (Cohen 1999: 267).

It is possible that the absence of Canaanean blade technology is related to the lack of intensive agriculture in the desert regions (for a comparison between the sickle blade frequency and the Canaanean blades distribution see Rosen 1997: 58-9, figs 3.17-3.18; Baird 2001: 648). The locally made retouched blades and the arched backed blades used as reaping implements were probably adequate for the opportunistic agriculture practiced by the desert inhabitants.

Bead production constitutes the primary case for a certain kind of specialized production and exchange in the Early Bronze Age IV in relation to stone tools. The cache of beads and 36 microlithic drills found in locus 22 in area B at Horvath Beer Resisim reflects a specialized cottage industry. The *Chaîne opératoire* of bead production is actually complex. On one hand, we may trace the drill manufacturing sequence from raw material, through the production of cores and blanks, and completed with the production of the drills, their use, and discard (Roshwalb 1981). On the other hand, there must be a parallel process in the procurement and preparation of the bead materials and beads, some deriving from as far as the Red Sea in the case of seashells, and South Sinai in the case of minerals. Local materials, like ostrich eggshells may also have been available (Bar-Yosef 1999: 322-5). The complexity and distances reflected in this cottage industry contrast with the simplicity of the bulk of the lithic industry at Beer Resisim, and indicate a social and economic organization beyond simple domestic consumption and production. Although the ultimate consumers of the beads produced at Beer Resisim remain unknown, it is possible that these items were traded both with close and distant regions. The long distance between the site and the raw material source certainly indicates that some value was attached to these exotic items. Since no cities existed during the Early Bronze IV, it is difficult to pinpoint potential markets, and it is possible that the beads served primarily within Early Bronze IV society as some kind of social marker.

Similar small bead making ateliers have been documented at the Camel site (Rosen 1994-5) and at the site of Nahal Tzafit. Both sites date to the Early Bronze I -II. In all these three cases, finished beads and raw materials were found together in proximity to microlithic drills and drill manufacturing waste (Rosen 2003, Vardi *et al.* 2003). The presence of such small-scale bead ateliers at Early Bronze I-II and Early Bronze IV sites suggests some level of basic economic continuity between the two periods.

The question of agriculture

The low frequencies of sickle blades at the sites of Beer Resisim and Ein Ziq might hint to a very limited (if at all existent) agriculture. The total number of sickles from Beer Resisim (n=22) and from Ein Ziq (n=5) is minimal. In this context, it is important to note that sickle gloss can be the outcome of reaping or cutting of different kinds of plants, not only of harvest cereals (Anderson 1980). Given the presence of numerous grinding stones, and obviously a relatively large population, it is likely that grain was imported to these sites from the Judean hills, the Shefela or from the northern Negev. It is very unlikely that the local arid small-scale agriculture could have supported the large number of residents occupying the central sites. Comparison to assemblages from northern Early Bronze IV sites (Dar 1977, Rosen 1983, 1997:129) reveals a very high ratio of sickles compared to the south. This difference reflects above all a basic difference between two modes of subsistence, the northern one relying on agriculture and the southern one a pastoral economy, relying on herd management and trade.

The abundance of hammer stones, large baseball-sized pecked cobbles, found at these sites, also requires comment. Although the hammers form only a fraction of the assemblages both at Ein Ziq and at Beer Resisim, their numbers are large in comparison to Early Bronze I-II assemblages from other sites in the Negev Highlands. Only ten hammer stones were found at the Camel Site and eight at Nahal Tzafit. Furthermore, the majority of the hammer stones were not collected or were discarded by the excavators (Y. Israel *pers. comm.*). Cohen noted that each dwelling produced several hammer stones, usually in groups of three to eight, suggesting that the original assemblage contained hundreds of these items. Although hammer stones are often used in flint knapping, Haiman (1992: 128-9, 154, 1996) and Cohen (1999) have suggested that their high numbers reflect use as pounding stones for shaping copper objects in the Early Bronze IV metallurgical industry, represented by copper ingots and tools at these sites.

An amount of three to eight hammers from each dwelling (room) give a minimal number of 750 hammers for Ein Ziq (3 x 250 dwellings) and 300 for Beer Resisim (3 x 100 dwellings, see Haiman 1992: 157, Tables 2-3 for architectural data). This enormous amount rather suggests to us an intensive use of hammers as pounding stones for domestic use, such as the crushing of various organic materials, firstly because the number of hammers correlates well with the number of grinding stones; and secondly, because there is no material evidence for intensive metallurgical activities at Ein Ziq and Beer Resisim (such as hearths, slags, crucibles, copper prills and other expected metallurgical products and by-products). We believe that Dever's (1995) description of the Early Bronze IV central sites' involvement in the trade of finalized copper products is more accurate.

Conclusions

The lithic industries of the third millennium BCE in the Negev show basic technological and functional continuity. The tool assemblages from Beer Resisim and Ein Ziq are similar to earlier Early Bronze I-II assemblages from the Negev Highlands, both typologically and in their basic quantitative configurations. Technologically, the basic reduction sequences, with the emphasis on *ad hoc* production, are also similar. The absence of Mediterranean elements, like the Canaanean blade, can also be interpreted as part of this general continuity. The primary differences lie in the absence of arrowheads and the virtual absence of tabular scrapers in the later period, and these seem to reflect variations in specific functions, as opposed to fundamental lithic changes.

Acknowledgments

This study was conducted thanks to the support of the National Science foundation (Israel); and as a part of the "Analysis of Third Millennia BCE flint Assemblages from the Negev Highlands Project" headed by Prof. Steve Rosen.

We would like to thank Mr. Benjamin Saidel and Mrs. Eveline van der Steen for their kind invitation to present this paper at the "On the Fringe of Society" seminar. We would also like to thank Alter Fogel who photographed the flint tools from Beer Resisim.

Bibliography

Aharoni, Y.
1978 *The Archaeology of the land of Israel*; translated by A.F. Rainey. Philadelphia: Westminster.

Amiran, R.
1978 *Early Arad : the Chalcolitic settlement and Early Bronze city*. Jerusalem: Israel Exploration Society

Anderson, P. C.
1980 "A Testimony of Prehistoric Tasks: Diagnostic Residues on Stone Tool Working Edges". *World Archaeology* 11: 181-194.

Andrews, C.
1990 *Ancient Egyptian Jewellery*. London: British Museum.

Avner, U.; Carmi, I.; and Segal, D.
1994 "Neolithic to Bronze Age Settlement of the Negev and Sinai in Light of Radiocarbon Datings: A View from the Southern Negev" in *Late Quaternary Chronology and Paleoclimates of the Eastern Mediterranean*, ed. O. Bar-Yosef., and R.S. Kra. Tuscon: University of Arizona. pp. 265-300.

Baird, D.
1987 "A Preliminary Analysis of the Chipped Stone from the 1985 Excavationes at Tell Esh-Shuna North". *Annual of the Department of Antiquity of Jordan* 31: 461-80.
2001 "The Analysis of Chipped Stone" in *The Archaeology of Jordan*, eds B. MacDonald; R.A. Adams; and P. Bienkowski. Sheffield: Sheffield Academic Press. pp. 639-51.

Bar-Yosef, D.
1999 "Shells from Three Middle Bronze Age I Sites in The Negev Highlands" in *Ancient Settlements of the Central Negev Vol. I.:The Chalcolitihc Period the Early Bronze Age and the Middle Bronze Age I, IAA Reports, No. 6*, ed. R. Cohen. Jerusalem: Israel Antiquities Authority. pp. 322-26.

Bar-Yosef, O.
1970 The Epi-Paleolithic Cultures of Palestine. Unpublished Ph.D Dissertation, Jerusalem: Hebrew University.

Betts, A.V.G.
1992 "The Chipped Stone Assemblage" in *Excavations at Tell Um Hammad 1982-1984: The Early Assemblages (EB I-II), Excavations and Explorations in the Hashemite Kingdom of Jordan*, ed. A. V. G. Betts. Edinburgh: Edinburgh University. pp. 122-131.

Bookman (Ken-Tor), R., Enzel, Y., Agnon, A. and Stein, M.
2004 "Late Holocene Lake Levels of the Dead Sea". *The Geological Society of America Bulletin* 116: 555-571.

Burian, F., and Friedmann, E.
1985 "Flint Borers: Methods of Production and use". *Mitekufat Haeven, Journal of the Israel Prehistoric Society* 18: 63-66.
1987 "Chalcolithic Borer Industry at site 103, Nahal Nitzana". *Mitekufat Haeven, Journal of the Israel Prehistoric Society* 20: 160-172.

Cohen, R.
1999 *Ancient Settlement of the Central Negev - vol. I. The Chalcolithic Period, the Early Bronze Age and the Middle Bronze Age I. IAA Reports, No. 6.* Jerusalem: Israel Antiquities Authority.

Cohen, R., and Dever, W. G.
1978 "Preliminary Report of the Pilot Season of the Central Negev Highlands Project". *Bulletin of the American Schools of Oriental Research* 232: 29-45.
1979 "Preliminary Report of the Second Season of the Negev Highlands Project". *Bulletin of the American Schools of Oriental Research* 236: 41-60.
1981a "Preliminary Report of the Third and Final Season of the Central Negev Highlands Project". *Bulletin of the American Schools of Oriental Research* 243: 57-77.
1981b "Beer Resisim". *Revue Biblique* 4: 571-573.

Crowfoot Payne, J.
1983 "The Flint Implements of Jerico" in *Jerico IV*, eds K. M. Kenyon and T. M. Holland. London: British School of Archaeology in Jerusalem. pp. 622-758.

Dar, S.
1977 *Ancient settlements of Emek-Hefer: The Sites of Tell Nurit, Nahal Alexander and the Rockcut Tombs of Maabarot*. Kibutz Maabarot.
Dever, W. G.
1970 "The 'Middle Bronze I' Period in Syria and Palestine" in *Near Eastern Archaeology in the Twentieth Century: Essays in Honor of Nelson Glueck*, ed. J. A. Sanders. New York: Doubleday. pp. 132-163.
1989 "The Collapse of the Urban Early Bronze Age in Palestine: Towards a Systemic Analysis" in *L'urbanisation de la Plaestine a l'age du Bronze Ancien, vol. I*, ed. P. de Miroschedji. Oxford: B.A.R International Series. pp. 225-246.
1995 "Social Structure in the Early Bronze IV Period in Palestine" in *The Archaeology of Society in the Holy Land*, ed. T. E. Levy. London: Leicester University. pp. 282-296.
Gilead, D.
1973 "Flint Tools of the Middle Bronze Age from Har Yeroham and Tell Nagila" in *Excavations and Studies* (Hebrew; English summary), ed. I. Aharoni,. Tel Aviv: Tel Aviv University. pp. 133-143.
Gilead, I.; Hershman, D.; and Marder, O.
1995 "The Flint Assemblage from Grar" in *Grar: A Chalcolithic Site in the Negev*, ed. S. Ahituv. Beer Sheva: Ben Gurion University of the Negev. pp. 223-281.
Glueck, N.
1968 *Rivers in the Desert: A History of the Negev*. New York: Norton.
Gophna, R.
1992 "The Intermediate Bronze Age" in *The Archaeology of Ancient Israel*, ed. A. Ben-Tor. New Haven: Yale University. pp. 126-59.
Goren, Y.
1996 "The Southern Levant in the Early Bronze Age IV: The Petrographic Perspective". *Bulletin of the American Schools of Oriental Research* 303:33-73.
Haiman, M.
1992 Early Settlement Patterns of the Negev Highlands: Analysis of the Findings of the Negev Emergency Survey 1979-1989. Unpublished Ph.D Dissertation. Jerusalem: Hebrew University.
1996 "Early Bronze Age IV Settlement Pattern of the Negev and Sinai Deserts: View from Small Marginal Temporary Sites". *Bulletin of the American Schools of Oriental Research* 303:1-33.
Hakker-Orion, D.
1977 "The Animal Remains of Nahal Alexander Site" in *Ancient settlements of Emek-Hefer: the Sites of Tell Nurit, Nahal Alexander and the Rockcut Tombs of Maabarot*, ed. S. Dar. Kibutz Maabarot. pp. 92-101.
Hughes-Brock, H.
1998 "Greek Beads in the Mycenaean Period (ca 1650-1100 BC): The Age of Heroines of Greek Tradition and Mythology" in *Beads and Bead Makers: Gender, Material Culture and Meaning*, eds L.D. Sciama, and J.B. Eicher. Oxford: Berg. pp. 247-72.
Idelman, A.; Kaplan, M.; Salmon, A.; and Enmar, A.
2000 "The Negev Lithology" in *Landscape Units in the Negev* (Hebrew), eds A. Idelman; M. Kaplan; and A. Sadot. Jerusalem: The Israel Nature and National Parks Authority. pp. 29-33.
Inizan, M.-L.; Roche, H.; and Tixier, J.
1992 *Technology of Knapped Stone*. Meudon: Centre National de la Récherche Scientifique.
Inizan, M.-L., Reduron-Ballinger, M., Roche, H. and Tixier, J.
1999 *Technology and Terminology of Knapped Stone*. Nanterre: Groupe de Recherches d'Études Politiques (CREP).
Kenyon, K. M.
1960 *Archaeology in the Holy Land*. London: Benn.
Kozloff, B.
1972-3 "A Brief Note on the Lithic Industries of Sinai". *Museum Ha'aretz Yearbook* 15/16:35-49.
Leakey, M.
1971 *Olduvai Gorge, Vol. III*. Cambridge: Cambridge University.
McCartney, C.
1996 "A Report on the Chipped Stone Assemblage from Tell Iktanu, Jordan". *Levant* 28:131-154.
McConaughy, M.
1979 Formal and functional analyses of the chipped stone tools from Bab edh-Dhra, Jordan. Unpublished Ph.D Dissertation. Pittsburgh: University of Pittsburgh.
1980 "Chipped Stone Tools [from Bab edh-Drah]". *Bulletin of the American Schools of Oriental Research* 240:53-58.
Miller-Rosen, A.
1995 "The Social Response to Environmental Change in Early Bronze Age Canaan". *Journal of Anthropological Archaeology* 14:26-44.
Neev, D., and Emery, K. O.
1995 *The Destruction of Sodom, Gomorrah and Jericho: Geological, Climatological and Archaeological Background*. Oxford: Oxford University.
Neuville, R.
1934 "L'outillage en Silex" in *Teleilat Ghassul I*, eds A. Mallon; R. Koeppel; and R. Neuville. Rome: Institute Biblique Pontifical. pp. 55-65.
Odell, G. H.
2000 "Stone Tool Research at the End of the Millenium: Procurement and Technology". *Journal of Archaeological Research* 8: 269-324.
Palumbo, G.
1991 *The Early Bronze Age IV in the Levant: Settlement Patterns, Economy, and Material Culture of a 'Dark Age'*. Contributi E Materiali di Archeologia Orientale, Vol. III. Rome: Universita Degli Studi di Roma.
2001 "The Early Bronze IV" in *The Archaeology of Jordan*, eds B. MacDonald, R.A. Adams and P. Bienkowski. Sheffield: Sheffield Academic Press. pp. 233-271.

Prag, K.
1974 "The Intermediate Early-Middle Bronze Age: An Interpretation of the Evidence from Transjordan, Syria and Lebanon". *Levant* 6: 69-116.

Rosen, S. A.
1982 "Flint Sickle Blades of the Late Protohistoric and Early Historic Periods in Israel". *Tel Aviv* 9:139-145.
1983 "The Canaanean Blade and the Early Bronze Age". *Israel Exploration Journal* 33: 15-29.
1990 "Appendix A: The Lithic Assemblage of Sites 911 and 915 at Biqat Üvda". *Atiqot* 10: 4*-9*.
1994-5 "Microlithic Drills from the Camel Site, Mizpe Ramon". *Journal of the Israel Prehistoric Society* 26: 148-158.
1997 *Lithics after the Stone Age*. Walnut Creek: Alta Mira Press.
1998 "The Chipped Stone Assemblage" in *Villages, Terraces and Stone Mounds: Excavations at Manahat, Jerusalem 1987-1989*, IAA Reports, No. 3, eds G. Edelstein; I. Milevski; and S. Aurant. Jerusalem: Israel Antiquities Authority. pp. 78-88.
2001 "The Lithic Assemblage from Üvda Valley site 917 and its Spatial Implications". *Atiqot* LXII: 109-119.
2003 "Early Multi Resource Nomadism: Excavations at the Camel Site in the Central Negev". *Antiquity* 77: 750-761.

Regional Markets and their Impact on Agriculture in Mamluk and Ottoman Transjordan

Bethany J. Walker

Abstract

The decline of village life in Jordan after the fifteenth century and a parallel resurgence of pastoralism, which are lamented in historical sources and suggested by regional surveys, have traditionally been attributed to political collapse and environmental change. While natural phenomena, such as temperature change and drought, are certainly important factors, financial documents of the period suggest that market pressures and peculiarities of the provincial administration had more of a long-term impact on village life and agriculture in general. Three modern villages in central and northern Jordan have been selected for multidisciplinary investigations of these factors within the rubrics of a political ecological model. This area of inquiry combines research methods culled from the disciplines of history, archaeology, anthropology, geography, and geology and makes use of technologies originally developed in microbiology and electrical engineering to reconstruct the totality of human behaviour that has transformed the environment.

This will be accomplished, in part, by identifying and quantifying the most archaeologically visible components of such cycles of exploitation and neglect: settlement and cropping patterns. The modern villages of Hisban in the Madaba Plains of central Jordan and Malka and Hubras in the Sawad of the north have been selected for this study because they are richly documented in contemporary written sources and have significant archaeological remains.

The identification and excavation of architecture and installations related to food processing and storage and village life as a whole, combined with water surveys, oral interviews with local landowners, and the extraction of medieval and modern soil samples, document the historical cycles of agricultural prosperity and decline and illustrate ways in which the villages have physically changed and their environments have been transformed since the 14th century in response to the market demands of lucrative export commodities, namely processed cane sugar and olive oil.

The Decline of Late Islamic village life – The "Desert vs. Sown" model discredited

Modern historical and archaeological scholarship on Jordan emphasizes a general decline in settlement during the 15th century AD. Regional surveys have consistently identified a reduction in the number of permanently settled villages and the abandonment of politically important centres in Transjordan, supporting "on the ground" the claims of contemporary Arabic sources that the political turmoil of the period created conditions detrimental to agricultural pursuits (Shawkat 2002) Until only recently, the low density of villages, as compared to earlier periods, has been attributed to the predations of local Bedouin tribes, which are blamed time and time again in Mamluk and Ottoman sources for the economic and social decline of Transjordan after the 15th century. The language of these accounts has resurfaced and been transformed by modern scholarship as an Orientalist critique of demographic transformations, which describe the Ottoman period as a whole as one of corruption, ignorance, violence, and official apathy. Central to this intellectual orientation is an assumption that farmers and pastoralists exist at opposite ends of a socio-economic spectrum and that their interests are always at odds with one another. Such critique also implies that agriculturalists are historical victims of imperial policies or tribal violence, rather than being actors themselves, who make their own informed and self-interested decisions. Following this line of logic, then, villages are abandoned each time they are vulnerable to attack by tribesmen and then resettled when security is restored.

It is beyond the scope of this paper to describe the complex and multiple-layered relationships among peasants and pastoralists in Jordanian society; my colleagues have dealt expertly with such topics through their presentations for this conference. I would, however, like to explore the issue of the decision-making process by farmers in the Late Islamic period by raising two questions: 1. to what extent did agricultural markets (and the financial incentives promised by them) impact cropping practices (choosing what to grow and how), and 2. can changes in these practices help us to understand settlement shifts in the Late Mamluk and Ottoman periods? The results of recent studies in Jordan on traditional agriculture and markets have identified the most important social, economic, and environmental factors behind modern farmers' decision-making. They provide the intellectual and methodological framework for research on pre-modern agricultural management.

Models for studying agricultural management

NGOs and ethnographic studies

There has been increased interest by non-governmental organizations since the 1980s on sustainable agriculture in Jordan. The focus of their development strategies has been, understandably, on water resources, but the topics of recent conferences share other themes: appropriate use of land, given certain soil types, slope, and rainfall; marketing strategies; and government intervention in cropping and regional and export markets (Qasem 1986; Taquieddin and Mitchell 1986; and 'Abdullah 1989). Throughout there is a concern for fragmentation of small land holdings and absentee ownership, which leaves 34% of agricultural land uncultivated each year; food imports, which consume most of the foreign currency earned by exports of non-food commodities; and long-term environmental impact, specifically the erosion and depletion of soils and misuse of limited water resources (Qasem 1986: 17; Taquieddin and Mitchell 1986: 128). The abandonment of wheat for vegetable farming (and particularly high-revenue tomatoes) best exemplifies all of these trends. The results of these studies suggest that both the cause and solution to these problems lies with government intervention in the areas of cropping and marketing.

Ethnographic studies on modern Jordanian agriculture have repeated the same themes. In her 1998 article on several villages in northern Jordan, Carol Palmer explores the ways in which local farmers using traditional technologies operate as part of the modern global economy (Palmer 1998: 161). In doing so, she argues that the economic rational for decision-making in traditional societies, that is the desire for immediate or dependable revenues, is balanced by how farmers conceptualize and store surplus, manage risk, and cope with scarcity. Outside of the use of fallow seasons and crop rotation, little consideration is given to the impact on the environment in making cropping decisions (Palmer 1998: 157). Similarly, in her analysis of the factors behind the Syrian grain boom of the late nineteenth century, Lynda Schilcher emphasizes the prerogatives of individual farmers and farming communities and how they measured benefits gained by land registration and export markets (Schilcher 1991). By adopting a model of supply-and-demand, she explains that certain conditions were in place to encourage farmers to plant wheat and export their surpluses: adequate land and labour, good climatic conditions, security and good transportation networks, organized production, regional diversification of agricultural production, and sufficient internal and external demand (Schilcher 1991: 181-185). Interestingly, in this model the farmer plays a more important role as decision-maker than the Ottoman state, which could not have privatized land through the 1858 Land Code without the support and active collaboration of the farmers.

While their interests do not always coincide, and there was historically much abuse of the peasantry by the state for financial gain, there has also always been a somewhat symbiotic relationship between the peasant and the state: neither acts entirely without the other, there is often considerable collaboration between the two, and the decisions of one produce a reaction by the other. Food systems theory examines all the human activities related to the acquisition, preparation, storage, and disposal of food. Øystein LaBianca's research on Late Ottoman Jordan has made use of food systems models to explain how village communities survive periods of official neglect and interference and how cycles of settlement intensity and abatement, archaeologically documented by regional surveys, can be related to culturally informed strategies for obtaining food (LaBianca 2000 and 1990).

All of these models suggest that the calculations made by farmers about where they will live and how they will use their land, while primarily economically driven, rely on a variety of factors, not the least of which are land tenure customs, infrastructure and security, labour supplies, and market knowledge (local and regional). Moreover, farmers have considerable autonomy in making residence, cropping, and marketing decisions. Whether this was also true for the Mamluk and Ottoman periods is suggested by the results of archaeological projects in northern Jordan and the Jordan Valley.

Environmental designs for archaeological projects

The Yarmouk and Jordan Rivers and their tributaries water the best agricultural land in Jordan. The Wadi Ziqlab survey of the 1980s, directed by Ted Banning of the University of Toronto, was designed to debunk the old "desert vs. sown" theories by investigating two related problems: to what degree pastoral-peasant relations determined cycles of settlement and what environmental variables and agricultural strategies affected the locations of villages (Banning 1982 and 1985). Such lines of inquiry necessitated a variety of data collection methods, including a thorough documentary study, pollen analysis, and geological and soil analysis. The results of the survey suggest that while the physical space that peasants and pastoralists exploit overlap, they do not degrade one another's environment and more often than not cooperate in resource management (Banning 1982: 4).

A more recent project by a British team in the Jordan Valley is a targeted investigation of the Mamluks' plantation economy. The excavation and geophysical survey of Tawahin al-Sukkar (a Mamluk sugar mill) and Khirbat al-Shakh 'Isa (a contemporary settlement nearby) in 2002 had as their immediate goals technological and environmental analyses of architecture, installations, soils, and water, in order "to place the sugar industry in the context of the landscape that generated and sustained it" (Photos-Jones *et al.* 2002: 591). The project made use of ground-penetrating radar (GPR) to locate underground installations related to the mill's use and to determine the physical parameters of the production site and mineral analysis of industrial waste to determine the process of sugar clarification. Their long-term objectives include determining the size and nature of the settlement that supported the mill, locating the transportation routes and markets related to the local sugar trade, describing the organization of the work force behind the mill's operation, and identifying the topographical and climatic conditions (such as slope, water resources, and soil types) required for successful sugar manufacture.

The research model adopted for my study was designed to isolate those factors modern NGOs examine in evaluating agricultural policies – namely markets, land tenure, and environmental impact – by adapting the data collection methods of environmentally focused archaeological projects. The relationship between settlement abatement in the Late Islamic period with shifts in land management strategies, however, begins with a thorough analysis of rural administration in the Middle Islamic period. This is accomplished through a careful reading of archival documents related to land use and published, primary sources of the period.

Agricultural markets in pre-modern Jordan – the documentary evidence

Mamluk period

Although the Mamluk period is one of the most richly documented in medieval Islam, contemporary written sources specifically related to land use and agriculture are rare. Tax registers, which provide us with the most information for the Ottoman period, no longer exist for Mamluk times, and, outside of a few examples for the region in the Haram archives in Jerusalem, writs of sale, purchase, and written testaments of rural real estate are seldom preserved.[1] What does remain, however, are endowment documents (*waqfiyyat*) describing rural estates, the revenues of which are set aside for social services; the so-called "secretaries' manuals" that guide state financial officers in provincial and rural administration; and chronicles that occasionally describe political and environmental events affecting local harvests and tax collection. In addition to these Arabic sources, contemporary mercantile letters and consular letters in European archives document the mechanisms behind the international exchange of agricultural produce.

In order to assess the overall impact of the Mamluks' agricultural policies on settlement and rural – Bedouin relations, several questions need to be asked of these sources, limited as they are:

1. What were the Mamluk agricultural policies? Did Egypt dictate what would be grown in the more distant provinces? In what ways did they succeed or fail? What was their long-term impact?
2. Was land fragmentation (the result of dispersed *iqta'at*) a problem? Did it result, for example in reduced grain yields?
3. What was the structure of agricultural markets and marketing in Transjordan? (How did it compare to other regions of the empire?) Did the state in any way control these? Where were the largest agricultural markets in Transjordan? How did they operate?

Although the written sources at our disposal give us precious little information about the Mamluks' administration of rural Syria, a recent monograph by Tsugitaka Sato (1997) describes how the state interacted with the peasantry in Egypt. Relying largely on the accounts of the management and planting cycles of Egyptian agricultural land by al-Makhzumi, Ibn Mammati, al-Nabulsi, and al-Nuweiri, he describes the numerous bureaucrats sent to the villages to record annually details about the land, reassign planting certificates, and distributing grain, where necessary. Nonetheless, these representatives of the state and the Mamluk officers ('muqta's') to whom were assigned the revenues from this land rarely interfered in the internal operations of the planting and harvests. It was the peasants themselves who decided what to plant, on what schedule to rotate crops, and how to share water. In other words, local custom generally prevailed in matters of cropping, harvest, and processing. One notable exception is sugar production, which was more closely monitored by the muqta', who tended to be the sultan himself. On "sugar estates" the cropping of the sugar plant took precedent over other crops and customary water sharing agreements, interrupting crop rotation and the planting of summer crops (Sato 1997: 212, 233). There is, unfortunately, no comparable source material about the management of agricultural lands in Transjordan. The image we do get from random references in Syrian chronicles and administrative manuals is that the Mamluk state had no agricultural "policy" of which to speak in the region. Its administration of the southern *Bilad al-Sham* was uneven in application and constantly changing in structure in order to meet short-term political objectives (Walker 2003). We can, nevertheless, make a few generalizations, drawn from alternative documentation, about the general development of agricultural production and marketing in Transjordan. While the state was not consistent or far-sighted in its administration of its southern Syrian provinces, it was actively involved, at times, in certain sectors of rural infrastructure that benefited the local agricultural sector. Cairo invested in Jordan's communications network from the 13th century, by levelling roads, building military outposts and caravanserais, and reorganizing the postal (*barid*) and pigeon routes (Walker 2004a). The primary objective was, of course, defensive; however, such facilities also facilitated the transport of agricultural goods from fields to markets and on to ports. Early Mamluk sultans founded new villages in Egypt and Syria and revived long abandoned ones (Ashtor 1976: 316). Rural land throughout the empire was organized by the *iqta'* system, which required the military officer holding the tax collection rights of this land (the *muqta'*) to repair canals and dams and contribute to capital improvements of agricultural facilities. Such investments were beneficial to both peasants and the *muqta'*, who could enjoy higher profits through increased tax revenues.

Perhaps the most important administrative development, with regards to Jordanian agriculture, was the cadastral survey (*rawk*) of southern Syria by Sultan al-Nasir Muhammad (3rd reign 1310-1341 C.E.) in 1313-1325 (Walker n.d.[b]). The survey was the basis of reallocating landed estates (*iqta'at*) to military officers

[1] A catalogue of the Mamluk documents in this collection, largely fourteenth-century in date, can be found in Little 1984.

and resulted in breaking up officers' estates into smaller portions, distributed at distances from each other throughout the empire; concentrating more land in the hands of the sultan; and providing the financial means for the promotion of second-class soldiers (including *awlad al-nas*). The best of these *iqta'at* were clearly tied to export agriculture, as many consisted of plantation-like estates, were located on major transportation routes, and were equipped with extensive water and agricultural surplus storage facilities (Walker 2003: 246).

Some of the most productive agricultural lands became privately owned estates (*amlak*) through purchase from the Treasury by mid-century (Ashtor 1976: 318). The last quarter of the 14th century witnessed the endowment of many of these larger estates in Jordan (largely in the Jordan Valley and the wheat and olive-producing centres) to financially support the building and maintenance of *madrasas* (Islamic law schools) in Cairo, Jerusalem, and Damascus (Walker 2004c). This flurry of endowments coincided with heightened financial and political insecurity in Cairo and may have been an attempt by officers to protect their assets from confiscation (Walker n.d. [b] and 2004b). In short, Mamluk rural policies in this region were largely politically motivated and did not consider what was best for local markets (opposed to imperial markets) for the long term or for the environment. Sugar cultivation, while unquestionably profitable for the state, was especially hard on soil, depleting it of mineral resources and water (Photos-Jones 2002). Sugar cane must be planted on land that has been left fallow for at least four years (Sato 1997: 216). After two years' of harvest, the land must once again be left fallow or crops other than sugar should be planted. According to Ibn Mammati, land on which sugar was planted was taxed at a lower rate after the first year, thus losing value over time (Rabie 1972: 75).

Jordanian farmers experienced mixed benefits from such initiatives. Most of the Transjordanian market centres – such as Irbid, 'Ajlun, and 'Aqaba – were entrepots, serving merely as a middlemen between local peasant and bedouin clients, or as stops on the *hajj* route, servicing pilgrims to Mecca (Ghawanmeh 1987). There was always a marked regionalism in Jordanian agriculture, in terms of production and marketing. The Umm Qeis region was known for its olives and pomegranates, for example; the Madaba Plains for its wheat and walnuts; the region around Kerak for its wheat and cheese; and the Jordan Valley for its fruits, vegetables, and sugar cane (Ghawanmeh 1982a and 1982b: 80; Palmer 1998: 157-159). With state investment in infrastructure, such regions could now enjoy more regular commerce with their respective markets: Irbid and Damascus, the Mediterranean coast, and further afield with Iraq and Iran; 'Aqaba with the Hijaz and Egypt (Ghawanmeh 1987). Selective state investment meant that certain regions of the Transjordanian countryside were developed specifically for long-distance or export markets: sugar (for Mediterranean ports), wheat (surpluses sent to Cyprus and further afield), and even saffron (for European markets).[2] By the 15th century the larger estates went into decline, the result of amiral rebellions, official neglect, and changes in international markets (Walker 2004c). Many of these regions were abandoned, by both the state and villagers: parts of the Jordan Valley and the lowlands and plains of southern and central Jordan, for example. Even certain commodities in the past exported to Europe, most notably grain and olive oil, were now imported to Egypt and Syria from Italy, Spain, and the Dalmatian coast (Ashtor 1976: 319 and 1983: 222 and 465).

The "grain slump" from the mid-14th century on has been used by many historians as an index for agricultural decline in certain sectors and general economic malaise for the empire as a whole. This phenomenon is characterized by reduction in cultivated land, reduced grain prices (related to reduced demand), and imports of wheat and barley from abroad by the 15th century (Ashtor 1976: 315 ff and 1983: 435). While there has been no comprehensive study to date on this topic, I would suggest it resulted, in part, from the plagues that reduced Syria's rural population over the course of the 14th century, combined with some negative consequences of al-Nasir Muhammad's cadastral survey, which fragmented rural holdings, cutting lands off from natural water sources, separating villages from common markets, and making transport of agricultural goods to markets more difficult. Combined with this in the 15th century were governmental practices that interfered with the natural workings of the marketplaces and exacerbated market decline: forced purchases of state-controlled commodities (*tarh*), such as sugar and wheat; the hoarding of grains to artificially raise prices; and confiscations of goods and property.

Agricultural decline was not universal, however. While certain markets collapsed, others survived. Those regions not dependant on state support, that were already heavily populated, relied on local and regional (as opposed to foreign) markets, and enjoyed a diversified agricultural base survived the troubled times of the late Mamluk period; they were, in short self-sufficient (Walker 2004c). The districts of northern Jordan best exemplify this trend. Here indigenous farming and cropping traditions were quite similar to today. Local peasants observed a two-crop rotation and fallow seasons and used the scratch plow (*ard*) driven by two oxen or donkeys, practices that have been maintained by farmers in the northern districts into modern times (Ghawanmeh 1982b: 81, citing al-Umari).

Ottoman period

Ottoman bureaucracy was even more developed and extensive than that of the Mamluk state. Surveys and tax

[2] The literature on the Mamluk sugar industry is extensive. For a review of scholarship on this topic, see Walker 2003: 258-259 and 2004b. Saffron for export was produced in the village of Jadiyya in the Balqa (Ghawanmeh 1982b: 81).

registers (*defters*) of the 16th century give detailed information about land tenure, production and yield, and even markets for the Transjordanian districts, reflecting the structure of the late Mamluk period, which was largely adopted by the Ottomans early in their reign (Bakhit 1989; Bakhit and Hmoud 1989 and 1991; Bakhit *et al.* 1989). For the 19th century, we have at our disposal traveller's accounts; land registers, after 1880, and court documents related to property exchange (Bakhit *et al.* 1984); and consular records of exports. Such official records are supplemented by family documents (such as letters and receipts – the basis of Abujaber 1989) and ethnographic interviews of elderly villagers, which has become a popular technique for acquiring economic and social data on the late Ottoman and Mandate periods. Unfortunately, for Jordan there is practically no written documentation for the period between 1600 and 1800 AD. Information about agricultural markets for this period must rely on archaeological fieldwork and object analysis, which are hampered by our currently limited knowledge of Ottoman ceramic typology.

Most of what we believe we understand about agricultural markets in Ottoman Jordan is true only for the 19th century, during the richly documented Tanzimat period. For the first time in nearly three hundred years the state became an active participant in the local economic and political sphere, intervening directly in matters of land tenure, taxation, and administration. The legal act with the most significant local impact was the Land Law of 1858, first applied to 'Ajlun in 1876, which required individuals to register land and pay tax on it.[3] The purpose of the legislation was to tax all productive land in the province. Its application throughout Jordan, however, differed from region to region, adapting to local traditions of land ownership (Rogan 1999: 83). The state then invested in critical infrastructure by paving roads, building bridges, and providing security. In contrast to the Mamluks, such initiatives were economically driven from the beginning, aiming at increasing tax revenues and supporting agricultural markets. In order to extend the area of cultivation, the state then encouraged immigration to sparsely populated areas of Transjordan. Newly arrived Circassians, Palestinians, and Syrians settled in old, established towns (such as al-Salt) and created new ones ('Amman, for example); became a new entrepreneurial elite with trade ties to Damascus, Nablus, and Jaffa; and created a new consumer market (Abujaber 1989: 131-240; Rogan 1999: 70-82, 99).

Significant changes in land tenure and commercialization of agriculture were the results of such developments. Privatization of land became the norm, although the form that it took differed from region to region; privately owned, small family plots or small shares of communally-owned lands characterized northern Jordan, for example (the legacy of registration of *musha'* land), while "tribal plantations" were more common in the central plains (Rogan 1999: 84 and 89; see also Schilcher 1991: 193). At the same time, a series of political and economic crises in Europe (the Napoleonic wars, the Crimean War, the European Great Depression) created a demand for cheap grain; the mid-century "Mediterranean grain trade boom" started a process of agricultural commercialization that took advantage of these developments in transportation, new forms of land tenure, and larger labour resources (Schilcher 1991: 193; Rogan 1999: 99). More merchants from throughout Greater Syria were attracted to Transjordan by the higher prices charged there on agricultural commodities (Rogan 1999: 29). Over the course of the century Jordan went from a focus on local and regional markets, to a general export orientation, with more than one-half of all of Jordan's agricultural products exported by the 1890s (Rogan 1999: 113, 116). Wool from 'Ajlun and al-Salt was sent to Damascus for transport to France and the United States; grains from the Balqa were carried by camel caravan to Jerusalem and Nablus to Jaffa and on to distant Mediterranean ports; raisins from al-Salt entered European markets via Jaffa (Schilcher 1991; Rogan 1999: 113-114).

This is not to say that export trade replaced regional markets. The regionalism of Jordanian markets remained strong, with local peasants and Bedouin trading animal products (butter, wool) and specialized agricultural products (such as olive oil) for housewares, textiles, and imported goods. Traditional trading partnerships – 'Ajlun and Nazareth, Irbid and Nazareth or Haifa, Salt and Nablus, Kerak and Hebron or Jerusalem, Ma'an and Gaza – guaranteed that villages would not be completely subjugated to the capriciousness of international markets or state policies (Abujaber 1989: 103-104; Rogan 1999: 35).

Archaeological investigations on markets

The written sources on which this narrative is based suggest the political and economic agendas for agricultural initiatives by the government. They do not, however, fully explain the cycles of sedentarization and nomadization that seem to characterize the period or the role of peasants in market management. In an effort to better understand the factors behind these cycles and identify the relationship, if any, between these markets and settlement, I will very briefly summarize the preliminary results of a recent archaeological work at Tall Hisban, Khirbet Malka, and "Old Hubras" (Fig. 1).

MPP-Tall Hisban Excavations – The 1998, 2001, and 2004 seasons

The village of Hisban occupies a strategic position on the Madaba Plains, situated on the caravan and pigeon routes of the Mamluk period and astride an important road to Jericho and Jerusalem (Walker 2003: 249). Contemporary sources refer to the village as a centre of agriculture,

[3] Further registration of land was conducted by the Mandate authorities in the 1930s (Palmer 1998: 129-130).

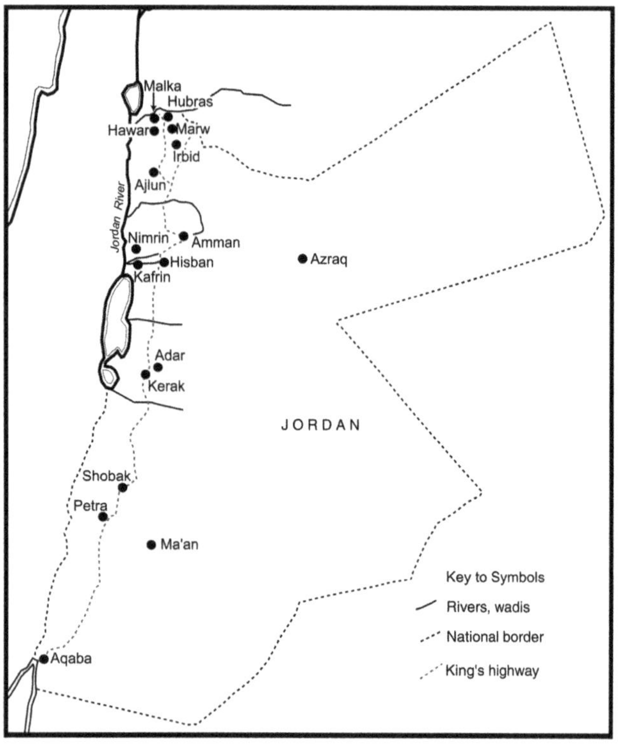

Fig. 1. Map of northern Jordan showing villages attested in written sources of Mamluk and Ottoman periods

Fig. 2. Wheat Fields of Wadi Majar, Hisban

controlling vast grain fields, gardens, and smaller villages, and boasting its own marketplace; the administrative capital of the Balqa in the first half of the 14th century; the home of many Qur'anic and hadith scholars who made their careers in Damascus, Jerusalem, and Cairo; and a tribal centre loyal to the Mamluk state (Walker 2001a, 2003, and 2004c – our Fig. 2). By the Ottoman period, it is barely mentioned. According to late 16th century tax registers the village was abandoned, although local tribes paid a fixed annual tax to the Ottoman state. While its ruins were used on a seasonal basis by the Ajarmeh and Adwan tribes, it seems to have not been historically documented again as a village until the 20th century (Hütteroth and Abdulfattah 1977: 169, entry P138).

Andrews University in Michigan began excavations at Tall Hisban in 1968, working at the site for several seasons until 1978. The Phase II project, conducted by a consortium of American and Canadian institutions and led by Andrews University, began in the 1998 season, with two subsequent seasons in 2001 and 2004, and is currently in the field. In addition to three historiographical articles (Walker 2001a, 2003, and 2004c), three preliminary reports from the Phase II operations have been published (LaBianca, Ray, and Walker 2000; Walker 2001b; and Walker and LaBianca 2003). On the basis of these reports the following observations can be made about Hisban's role in agricultural markets in the Mamluk and Ottoman periods.

1. The many cisterns that riddle the site, dated in one stage of use to the Mamluk period, attest to a concern for water storage. The Madaba Plain receives sufficient rainfall for dry farming of grains, so the vast water storage facilities may have been used for other industries requiring irrigation or water-processing, such as the refining of sugar cane. A storeroom full of glazed pottery and sugar pots from this period, in one wing of what the excavators identified as the Mamluk governor's residence, indicates that sugar was produced somewhere nearby. It has been suggested that Hisban served as a distribution point for sugar grown in the Jordan Valley, because no contemporary sugar mills have been positively identified during surveys of the wadis near the site. The *suq* described in written sources was probably a local marketplace, serving local villages and village-Bedouin exchange, while the citadel was oriented to regional trade.
2. The ceramic assemblage from the citadel storeroom combines local, handmade wares (primarily water jugs) with glazed relief ware bowls, likely produced in Jerusalem. Many of the latter bare Arabic inscriptions indicating that the vessels were made-to-order for military officials. There is also a significant quantity of glazed ware imports, the majority of them Egyptian. The pottery alone indicates official interest in local agricultural markets and the village's close ties with Cairo.
3. It is difficult to identify Ottoman remains at the site, given the poor typology of Ottoman pottery in Jordan, however several Bedouin cemeteries, caves in the surrounding hillside (once used for occupation and later as barns and storerooms), a series of buildings surrounding the tell, and the large *qasr* of the local Nabulsi family are likely Late Ottoman (Fig. 3).[4] The buildings are farmhouses, largely used for the storage of grains; additional surpluses could also be stored for

[4] This building complex has traditionally been called by the residents of Hisban a "qasr". Although the meaning of the term varies for time and place, in this context it is a farmhouse, used mainly to store fodder and farming equipment (Conrad 1981).

as long as three years in reused cisterns (Abujaber 1989: 103). Modern Hisban emerged with the registration of tribal lands and the commercialization of wheat production in the 1880s.

Fig. 3. The "Nabulsi qasr" in Hisban
The 19th-century farmhouse (in foreground) sits astride the southern approach to the tell (background).

Northern Jordan Survey – The 2003 season

In order to compare the effects of Mamluk rural administration in Hisban with other regions of the country, I began an environmental-archaeological project in northern Jordan last fall. Two villages were targeted for survey in this first field season: Malka and Hubras, located between Irbid and Umm Qeis.[5] There are currently three studies on this season slated for publication: two are historiographical (Walker n.d.[b] and 2004c) and the other is a preliminary report (Walker n.d.[a]). Because of the agricultural importance of this region in the Mamluk and Ottoman periods, both villages are amply attested in written sources, and most significantly in financial documents. Like Hisban, Hubras does not appear in Syrian chronicles until the 14th century, when it suddenly emerges as an important market village. Contemporary sources also refer to the scholars ("al-Malkawi") who made their careers in the most prestigious madrasas of Syria and Egypt. One unpublished *waqifiyya* in the Centre for Documents and Manuscripts (*Dar al-Watha'iq*) in Cairo describes in detail the village of Malka in the 1380s, when Sultan Barquq made his personal estate there, including its olive groves and its farmland, *waqf* for his new madrasa in Cairo (Walker 2004c).[6] It also describes, in general terms, the village mosque, as well as abandoned fields and derelict presses (presumably for olive oil). Ottoman tax registers of the 16th century attest to the continued economic viability of this region, and travellers' accounts of the 19th century also describe small, but financially secure, settlements (Bakhit *et al.* 1989; Walker, *op cit*).

Fig. 4. Medieval Olive Press in Malka

The trend towards mono-crop production (centred on olive oil) is apparent at Malka in the 14th century and further develops through the Ottoman period. Of the numerous caves, tombs, and food processing installations recorded on the western edge of the modern village, one feature, Cave 12, was directly related to this industry (Fig. 4). This underground, industrial complex is 13.6 m wide and 11 m deep, preserves the remains of six press levers and one basalt grinding stone, and is similar in structure and scale to similar olive oil plants found throughout the Galilee and Golan in the Byzantine through Ottoman periods. According to calculations made for similar weighted lever plants at Hellenistic Maresha, and based on population estimates from early Ottoman census records, six hectare of olive groves would have supported this single plant at Malka and could have produced 13,000-27,000 litres of olive oil annually. Of this amount, over 10,000 litres would have been surplus, exceeding the needs of local consumption and thus available for export (Walker n.d.[b]). The provenance of pottery collected from the interior surface of this cave and nearby installations, based on stylistic analysis and petrography, indicate northern (Damascus and the Golan) and western (Mediterranean coast) orientations in trade for the 14th century (Holzweg n.d.).

Fig. 5. Historical Mosques of Hubras
Remains of the qibla wall and central mihrab of the Mamluk mosque are visible to the right of the 19th-century ruins.

[5] Previous surveys in the region lacked chronological clarity and were not intensive (Mittmann 1970; Schumacher 1890; and Steuernagel 1924-1926).

[6] This important manuscript is currently being prepared for publication.

Our survey in the village of Hubras was designed to investigate the remains of two historical mosques (one built inside the other), previously attributed to the Mamluk (13th century) and Ottoman periods (‘Abaydat 1996: 14-24; Ghawanmeh 1986: 49-64; Meinecke 1992: 65, entry #43 – our Fig. 5). The architectural survey resulted in a down-dating of the smaller of the two mosques to the late 19th century and in taking earthen mortar samples for pollen analysis from both structures. A marked reduction in the size of the congregation is indicated by the construction of the smaller prayer hall. The preliminary results of pollen analysis confirm the gradual abandonment of land, overgrazing, and eventual diversification of agriculture from the 13th to 20th centuries (Cordova n.d.).

Conclusions – Markets and peasant-Bedouin relations in Ottoman Jordan

The residents of Hisban, Malka, and Hubras today continue to make profit-driven decisions about their land by considering the capabilities of local markets, their potential role(s) of their farms in global markets, the availability of government financial incentives, and employment opportunities outside of agriculture. Hisbanis and Malkawis have for years sent their sons into the armed forces for employment, making the sale of produce and grain a second income for the family. In Malka this has meant agricultural specialization, with an almost exclusive investment in olive oil and dairy products; the income earned from state jobs allows families to purchase other foodstuffs from the marketplace. The much smaller, and now poorer, Hubras has diversified its production in a way that it arguably did not do in the period under discussion. More residents here rely on their farms to feed their families, because fewer of them are employed outside of the village.

Clearly, today, as in the past, land use cannot be separated from expectations of market returns. Official agricultural policies and indigenous cropping practices, as they relate to these markets, impact the size, location, and longevity of settlements. The implications of this for the archaeological study of settlement and peasant-pastoral relations are significant. Future research on the decline of villages and apparent “gaps” in occupation in the pre-modern periods should look beyond the “desert vs. sown” models of the past and consider the decisions made by farmers about their own land.

Bibliography

‘Abaydat, M.
1996 *Dirasah fi Tarikh Baladah Hibras*. Irbid: Yarmouk University.
‘Abdullah, ‘A.R.A.
1989 *Jordan Environmental Profile – Status and Abatement*. Amman.
Abujaber, R.S.
1989 *Pioneers over Jordan: The Frontiers of Settlement in Transjordan, 1850-1914*. London: Taurus.
Ashtor, E.
1976 *A Social and Economic History of the Near East in the Middle Ages*. Los Angeles, CA: University of California.
1983 *Levant Trade in the Later Middle Ages*. Princeton, NJ: Princeton University.
Bakhit, M.A.
1989 *Tapu Defteri No: 275. Detailed Register of the Private-Khass of the Governor in the Province of Damascus 958 A.H./1551-2 A.D.* Amman.
Bakhit, M.A. and N.R. Hmoud
1989 *The Detailed Defter of Liwa' ‘Ajlun (The District of ‘Ajlun). Tapu Defteri No. 970, Istanbul.* Amman: University of Jordan.
1991 *The Detailed Defter of Liwa' ‘Ajlun (The District of ‘Ajlun). Tapu Defteri No.185, Ankara 1005 A.H./1596 A.D.* Amman: University of Jordan.
Bakhit, M.A. *et al.*
1984 *Kashaf al-Hisa'i Zamani li-Sijillat al-Mahakim al-Shari‘iyyah wa al-Awqaf al-Islamiyyah fi Bilad al-Sham, pt. I.* Amman: University of Jordan.
1989 *Nahiyat Bani Kinanah (Shamali al-Urdunn fi al-Qarn al-‘Âshir al-Hijri/al-Sâdi al-Milâdi.* Amman: University of Jordan.
Banning, E.B.
1982 “The Research Design of the Wadi Ziqlab Survey, 1981”. *Newsletter of the American Schools of Oriental Research* 8.8: 4-8.
1985 Pastoral and Agricultural Land Use in the Wadi Ziqlab, Jordan: An Archaeological and Ecological Survey. Unpublished Ph.D. dissertation, University of Toronto.
Conrad, L.E.
1981 “The Qusur of Medieval Islam: Some Implications for the Social History of the Near East”. *Al-Abhath* 29: 7-23.
Cordova, C.
n.d. Pollen Samples from Malka, Jordan: Preliminary Report. Unpublished report of 2004, Oklahoma State University files.
Ghawanmeh, Y.
1982a “al-Qarya fi Junub al-Sham (al-Urdunn wa-Filistin) fi al-‘Asr al-Mamluki fi Daw' Waqfiyyat Adar”. *Studies in the History and Archaeology of Jordan* 1: 363-371.
1982b *al-Tarikh al-Hadari li-Sharqi al-Urdunn fi al-‘Asr al-Mamluki*. Amman: Dar al-Fikr li al-Nashr wa al-Tawzi‘.
1986 *Madinah Irbid fi al-‘Asr al-Islamiyy*. Irbid: Marakaz al-Dirasat al-Urdunniyah.
1987 “Al-Tijarah al-Dawliyyah fi al-Urdunn fi al-‘Asr al-Mamluki”. *Studies in the History and Archaeology of Jordan* 3: 323-331.
Holzweg, L.
n.d. Petrography and Stylistic Analysis of the Middle Islamic Wares in the OSU Tall Hisban Study Collection. Unpublished B.A. senior honors thesis, Oklahoma State University.
Hütteroth, W.-D. and K. Abdulfattah
1977 *Historical Geography of Palestine, Transjordan and Southern Syria in the Late 16th Century.*

Erlangen: Selbstverlag der Fränkischen Geographischen Gesellschaft.

LaBianca, Ø.S.

2000 "Daily Life in the Shadow of Empire: A Food Systems Approach to the Archaeology of the Ottoman Period" in *A Historical Archaeology of the Ottoman Empire: Breaking New Ground*, ed. U. Baram and L. Carroll. NewYork: Kluwer Academic/Plenum Publishers. pp. 203- 217.

1990 *Sedentarization and Nomadization: Food Systems Cycles at Hesban and Vicinity in Transjordan.* Hesban 1. Berrien Springs, MI: Andrews University.

LaBianca, Ø.S., P.J. Ray. Jr., and B.J. Walker

2000 "Madaba Plains Project: Tall Hisban, 1998". *Andrews University Seminary Studies* 38.1: 9-21.

Little, D.

1984 *A Catalogue of the Islamic Documents from al-Haram aš-Šarif in Jerusalem.* Wiesbaden: Steiner.

Meinecke, M.

1992 *Mamlukische Architektur in Ägypten und Syrien (648/1250 bis 923/1517), pt. II: Chronologische Liste der Mamlukischen Baumassnahmen.* Glückstadt: Augustin.

Mittmann, S.

1970 *Beiträge zur siedlungs- und Territorialgeschichte des nördlichen Ostjordanlandes.* Wiesbaden: Otto Harrassowitz.

Palmer, C.

1998 "'Following the plough': the Agricultural Environment of Northern Jordan". *Levant* 30: 129-165.

Photos-Jones, E., K.D. Politis, H.F. James, A.J. Hall, R.E. Jones, and J. Hamer

2002 "The Sugar Industry in the Southern Jordan Valley: An Interim Report on the Pilot Season of Excavations, Geophysical and Geological Surveys at Tawahin as-Sukkar and Khirbat ash-Shaykh 'Isa, in Ghawr as-Safi". *Annual of the Department of Antiquities of Jordan* 46: 591-614.

Qasem, S.

1986 "Agricultural Land Use Policies" in *Agricultural Policies in Jordan,* ed. A. Burrell. Atlantic Highlands, NJ: Ithaca. pp. 11-29.

Rabie, H.

1972 *The Financial System of Egypt: A.H. 564-741 A.D./A.D. 1169-1341.* London: Oxford University.

Rogan, E. L.

1999 *Frontiers of the State in the Late Ottoman Empire: Transjordan, 1850-1921.* Cambridge: Cambridge University.

Sato, T.

1997 *State and Rural Society in Medieval Islam: Sultans, Muta's and Fallahun.* Leiden: Brill.

Schilcher, L.

1991 "The Grain Economy of Late Ottoman Syria and the Issue of Large-Scale Commercialization" in *Landholding and Commercial Agriculture in the Middle East,* eds. Ç. Keyder and F. Tabak. Albany, NY: State University of New York. pp. 173-95.

Schumacher, G.

1890 "Northern 'Ajlun, 'Within the Decapolis'". London: Palestine Exploration Fund.

Shawkat, R.H.

2002 *al-Tarikh al-Siyasii li-Mintiqah Sharq al-Urdunn (min Junub al-Sham) fi 'Asr Dawlah al-Mamalik al-Thani.* Irbid: Mu'assasah Himadah li-Dirasat al-Jami'iyyah wa al-Nashr wa al-Tawzi'.

Steuernagel, C.

1924-1926 "Der 'Adschlun". *Zeitschrift des Deutschen Palästina Vereins* 47 (1924): 191-240; 48(1925): 1-144 and 201-392; 49 (1926): 1-167 and 273-303.

Taquieddin, N. and M. Mitchell

1986 "Agricultural Marketing" in *Agricultural Policy in Jordan,* ed. A. Burrell. Atlantic Highlands, NJ: Ithaca. pp. 108-21.

Walker, B.J.

n.d. (a) "The Malka-Hubras Survey: Archaeological Investigation of Mamluk Agricultural Policy. Paper presented at ASOR Annual Meeting, Atlanta, Nov. 18-23, 2003". To be submitted to the *Annual of the Department of Antiquities of Jordan* 49 [2005].

n.d. (b) "The Politics of Land Management in Medieval Islam: The Northern Jordan Survey, 2004". Paper presented at Ninth International Conference on the History and Archaeology of Jordan: "Cultural Interaction through the Ages", Petra, Jordan, May 24-28, 2004. *Studies in the History and Archaeology of Jordan* 9 2005.

2001a "Mamluk Administration of Transjordan: Recent Findings from Tall Hisban". *Al-'Asur al-Wusta* 13.2: 30-33.

2001b "The Late Ottoman Cemetery in Field L, Tall Hisban". *Bulletin of the American Schools of Oriental Research* 322: 47-65.

2003 "Mamluk Investment in Southern *Bilad al-Sham* in the Fourteenth Century: The Case of Hisban". *Journal of Near Eastern Studies* 62.4: 241-261.

2004a "The Globalizing Effects of *Hajj* in the Medieval and Modern Eras", in *Connectivity in Antiquity: Globalization as Long Term Historical Process.,* ed. Øystein S. LaBianca, San Diego, CA: Continuum. New Approaches to Anthropological Archaeology Series. (in print).

2004b "The Islamic Age", in *Hesban 11: The Pottery,* ed. Larry Herr, Berrien Springs, MI: Andrews University. (forthcoming).

2004c "Mamluk Investment in Transjordan: a 'Boom and Bust' Economy". *Mamluk Studies Review* 8.2: 1-29. (in print).

Walker, B.J. and Ø.S. LaBianca

2003 "The Islamic *Qusur* of Tall Hisban: Preliminary Report on the 1998 and 2001 Seasons", *Annual of the Department of Antiquities of Jordan* 47: 443-471.

BET AL-MALAHI

Yuval Yekutieli

This paper is dedicated to the memory of my good friend 'Id al-Turi.

Abstract

The paper presents a research focused on a stone structure built in the 1930s by Salman al-Malahi, a Sheikh of the al-Turi family, and on its close surroundings. This locality, which is situated a few kilometres north of Beer-Sheva, was examined through archaeological field survey, analysis of aerial photographs and interviews. The archaeology-oriented research highlights several aspects of life in that region in the past, and enables a reassessment of some theoretical issues within Levantine archaeology.

Fig. 1. Beit al-Malahi. A Drawing by Genia Manor

'Id al-Turi

Introduction

An abandoned structure is visible on the west side of highway 40, *ca.* 7 km north of Beer Sheva (Figs 1, 2). As one approaches the structure white pieces of cloth can be seen hanging on its walls, which indicate that it has a certain status of holiness. Upon entering this complex, one finds oneself within the ruins of a large courtyard structure delimited by a high wall on the south and east, and flanked by rows of rooms on the west and north. Further examination reveals remains of hearths within the courtyard, and a path demarcated by small fieldstones that leads from the courtyard entrance into the westernmost of the three rooms on the structure's northern side (Fig. 3).

This path leads one into a room that appears to be the focus of the building. A glance inside this room reveals a low-built, rectangular, white-washed platform, which seems to demarcate a tomb. Placed in a standing position around this whitewashed platform are a concentration of opened tin cans and bottles (Fig. 4). When one looks up, it appears that the tops of the walls are crowned with pieces of white cloth, some of which are mounted on short wooden poles, in a flag-like fashion. The immediate impression that comes to mind is that the structure is a holy shrine, possibly a sheikh's tomb.

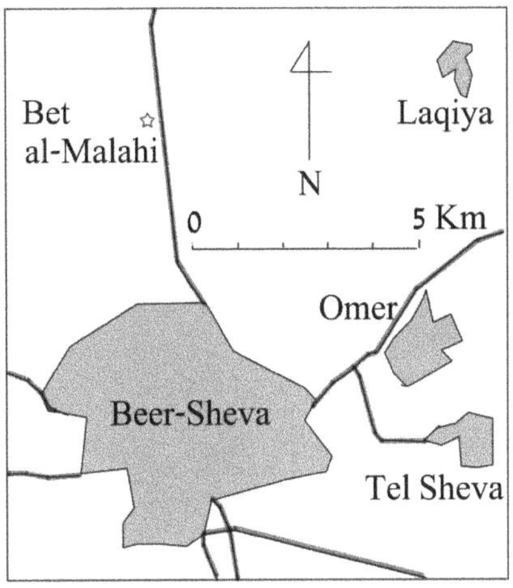

Fig. 2A.

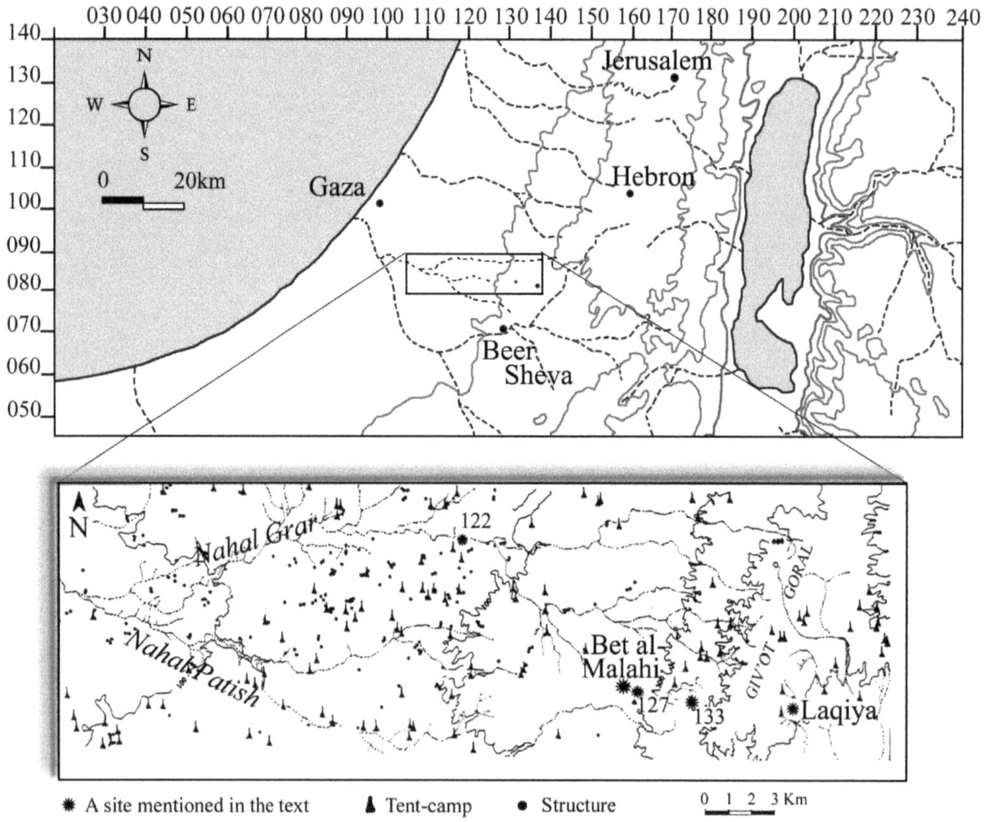

Fig. 2B

Fig. 2. A Location map, and an enlarged view of the research area, with indication of structures' and tent camps' locations in 1945 (according to aerial photographs)

Fig. 3. Within the courtyard of Bet al-Malahi, looking northwest. The entrance to room A in the centre, and to room C on the right

Fig. 4. The tomb in room C

The author's interest in this building, known locally as Bet al-Malahi, lead him and a group of colleagues in the early 1990s to investigate the history and function of this courtyard structure through an archaeological investigation, examination of aerial photographs, and an interview with one of its former residents (Yekutieli and Gabai 1995). The goals of this paper are to describe the results of that inquiry, and to discuss some archaeological implications of these results. The paper begins with a description of the geographical and historical context, continues with a presentation of the research results, and concludes with a theoretical discussion.

General Background

Geographical Setting

The structure known as Bet al-Malahi is situated on the border between two geographical regions: The "Northwestern Negev" with its flat to undulating Loess covered plateaus in the west, and the "Southern Shephela" characterized by rounded chalk hills in the east. Nahal Grar and its many tributaries drain the area westwards into the Mediterranean Sea (Fig. 2). Rains occur in the region from October to May. The study area is within the oscillation-range of the "aridity line" – which demarcates the geographical location of the 200mm precipitation line (Herzog 1990, fig. 2). This specific region is known for a unique climatic anomaly, whereby the average annual rainfall drops 4mm per kilometre as one moves south (Zangvil 1988: 46).

The deep Loess soils in the region are favourable for agriculture. Wheat may be grown in the area without irrigation, although the cultivation of vegetables and horticulture needs extra irrigation (Dan 1988: 56). In terms of natural flora, two physiographic zones are converging in this vicinity: the Mediterranean zone to the north and northwest, and the Irano-Turanian zone to the south. The region has known intensified herding within the last centuries, which have produced a selective pressure that only a few natural species have managed to survive (Danin 1988: 63).

Historical Context

The bedouin settlement system in the area under discussion "took off" during the last decades of Ottoman rule and especially during the period of the British Mandate of Palestine. This was mainly due to the region's location within a wider zone that played a major role in the struggle between the Ottoman and British Empires in the decades prior to and following the turn of the 19th century.

Events that testify to the growing interest of the Ottomans in the region during this period include the first founding of a stronghold within the bedouin territory at Qala't Futeis by the end of the 19th century, the establishment of Beer-Sheva as a central administrative centre in 1900, the introduction of a railroad into the area and various additional activities related to the proceedings of World War 1 in the region (Abu-Rabia 2001: 7).

After World War I, under British rule, the economy of the bedouin of the northern Negev witnessed a number of fluctuations. Following some prosperous years in the 1920s there was a general economic depression during the 1930s, which coincided with the global depression. Additional losses of income for the bedouin are attributed to the abolishment of *Ghazu* (robbery raids) and *Hawa* (protection payment). In addition the bedouin not only had to pay high taxes, but the value of their camels and horses had dropped owing to the overall transition to motorized vehicles. Furthermore, other traditional sources of income were impeded: quarrying of salt at Mount Sedom was prohibited, as well as its transportation without a permit; there was a ban on growing *Hishi* tobacco (which the bedouin grew during the Ottoman period) as well as its smoking beyond a restricted zone. Trade and possession of arms without a permit was outlawed, and to make things worse, a few drought years occurred in the Negev during the 1930s (Barslavski 1947: 137-139).

A positive change in the bedouin's economic situation occurred concurrently with World War II. Military preparations prior to and during the war were an economic boom for the bedouin and broadened their opportunities to participate in wage labour. For example, they earned money by working in military camps, which were located in the Negev. Some bedouin received large amounts of money as compensation for the confiscation of their land, which the British used to construct military camps and fortifications. With the outbreak of hostilities the British purchased sheep and goat from the bedouin in order to provide meat for their armed forces. Camels were also rented for the transportation of water and other loads. During the war the bedouin were able to command higher prices for their crops. The bedouin also profited from the sale of illegal and smuggled goods as well as stolen property from the British military. In sum, the war initiated activities and provided conditions that raised the standard of living of the Negev bedouin in general, and those of the Beer Sheva district in particular (Barslavski 1947: 140). In addition to improving the conditions of the local residents, the economic prosperity also attracted settlers from the Gaza area, who built a large number of buildings made of durable materials throughout the Northern Negev, thus changing the more common tent landscape of the region (David Alon, *pers. comm.*).

In the 1940s most of the area around Bet al-Malahi was within the territory of the Tiyaha confederation, of which the following tribes dwelt in the region: Ramadin, from Laqiya east and north; 'Arab Ibn 'Uqba in the area of today's Giv'ot Goral and Upper Karkur; 'Arab al-Huquq around Bir Fuhari; 'Arab en-Nutush near lower Karkur; and 'Arab abu-Medain and 'Arab abu-Dahar at the conjunction of Nahal Patish and Nahal Grar (Barslavski 1947: 247-249; the British Survey map at scale of 1:20,000, survey of Palestine 1947).

Results of the investigations

Archaeological Inquiries at Bet al-Malahi

The first step in studying Bet al-Malahi consisted of

mapping the structure and analyzing its architecture. Its good preservation as well as its general design made it clear that it was not built very long ago. Bet al-Malahi has a square plan, and measures 25 x 25 m. It is an enclosed structure, divided into an open courtyard, which comprises more than half of the entire complex; and a series of (once) roofed rooms, opening on to the enclosed courtyard along the western and northern sides of the structure (Figs 5, 6). On the southern and eastern sides of Bet al-Malahi is a perimeter wall that is preserved to a height of 2 m. Originally, the courtyard had a single opening on its eastern side.

Fig. 5. A 1993 aerial view of Bet al-Malahi

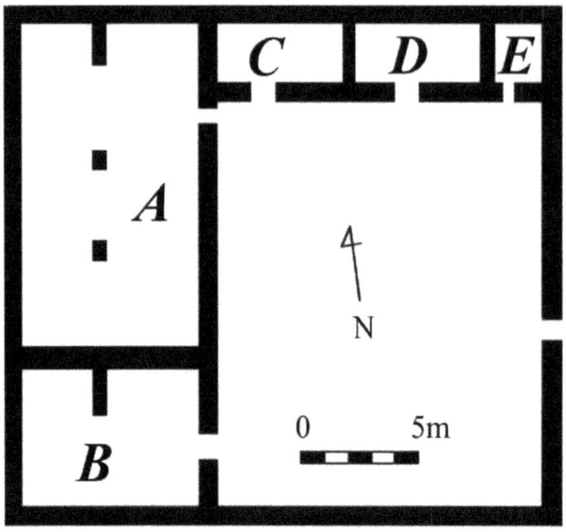

Fig. 6. Plan of Bet al-Malahi

The largest room, identified as Room A (Fig. 6), is situated on the structure's northwestern side. It is a rectangular room with two columns built on its central axis, each composed of plastered fieldstones, on which in the past a roof rested. Currently, one of the columns is still standing while the other has collapsed. A small pier is built against the northern wall on the same axis as the two columns. This room has a single opening into the courtyard. South of Room A, and occupying the entire remaining space towards the southern perimeter wall, stands room B, which is the second largest room in the Bet al-Malahi complex. This room has an inner pier constructed on the same axis as those in Room A. The northern end of the complex, to the east of Room A, consists of three rooms which all open into the courtyard. From west to east these three rooms are identified as rooms C, D, and E.

Starting at the entrance to Bet al-Malahi is a one metre wide path that crosses the courtyard and ends at the doorway of Room C. This walkway is delineated by two rows of small fieldstones (Fig. 5). Visible inside Room C is a low rectangular platform, apparently covering a tomb (Fig. 4). During our visits to the site we noticed that inside Room C there were many pieces of cloth, small open tin cans, small bottles, and lumps of incense wrapped in newspaper. Room D is a small rectangular space that opens on to the courtyard, while room E is the smallest room at Bet al-Malahi. Inside Room E, built into the western wall, there are many small niches arranged in two rows, one above the other. The niches are plastered with mud plaster that rounds their contours and narrows their openings to some extent (Fig. 7). The courtyard served as a central place into which all the rooms opened, and from which there was a way out of the structure. The remains of a few new hearths indicate some recent activity in the courtyard.

Fig. 7. The dove-cot at room E

The architectural analysis of the structure, and especially the way the walls are interrelated, either bonded or abutting each other, proves that room A was once a free standing structure – the first architectural unit constructed at Bet al-Malahi. Room B was later attached to the south side of room A (Fig. 8), and the row consisting of rooms C, D and E was joined to its northeastern corner (Note the wall bonding in Fig. 3). The courtyard's enclosing wall was built during the final architectural phase.

Fig. 8. Within the courtyard of Bet al-Malahi, looking west at the entrance to room B, and the association between the different walls

It seems that Bet al-Malahi was deserted after having been in use for a certain period of time. Despite the abandonment of the complex there are still activities being carried out inside Room C, where the tomb is located, and also in the courtyard, as borne out by the hearths.

An examination of the immediate periphery of the structure reveals large fieldstone heaps south of it, and two wide depressions dug in the Loess soil on its east (Fig. 5).

Interview with 'Id al-Turi

In order to acquire information on Bet al-Malahi and its inhabitants an interview was conducted with 'Id al-Turi of the al-Malahi tribe,[1] who had lived in this house in his early childhood. 'Id al-Turi told us that Bet al-Malahi was built by his grandfather, Salman al-Malahi, in the 1930s on a 1,000 dunam tract of land that he had purchased in that region. A master builder from the village of Dura (30 km to the northeast) was called in for the construction of the house. The large limestone blocks of which the structure is built, were brought on camel back from Hirbet Fuhari, an ancient ruin located approximately four km east of the construction site. Smaller stones were collected from the waste that accumulated from the quarrying of nearby water cisterns. Room A was built first, and its pier and columns served as partitions between spaces inhabited by each of the grandfather's wives. As the family grew in size room B was built and given to 'Id's father. In that room 'Id and several of his brothers were born. Rooms C, D and E were subsequently built together with the enclosing wall. Rooms C and D were used as dwelling spaces while room E was a dove-cot.

Regarding daily life in the structure, the rooms served as dwellings while activities such as cooking, hay storage, and keeping of the herd took place in the courtyard. The high wall surrounding the courtyard served to keep the herd together, and in particular, prevented thieves from observing what was going on within the household, while at the same time physically preventing them from entering it. At some point a guest hut (*Shig*) made of twigs and shrubs was built outside the structure against the exterior wall of room E. A passage was made in the wall of Room E, which lead directly into the guest hut. The stone piles located to the south of the structure had been collected as building material for a school that the grandfather had intended to build. East of the house large pits were dug into the Loess. These pits, which were once mud-plastered, were used to store grain, and sometimes to hide it from the tax collectors. On a certain occasion Salman al-Malahi planted a fig tree in one of them. For their subsistence the family cultivated some agricultural fields and maintained a small number of household animals. In addition it had many tents located away from the courtyard structure, where sheep and goats were raised, and later put up for sale.

According to 'Id al-Turi a series of consecutive events ultimately led to the abandonment of Room C as a dwelling and its subsequent reuse as a shrine. One of Salman al-Malahi's wives, an Egyptian woman who lived inside Room C, was unable to become pregnant. At approximately the same time contraband, such as certain types of tobacco, which were stored in this room, were found strewn in the courtyard outside it. On many occasions the inhabitants of Bet al-Malahi saw an image of an old man emerging and disappearing from Room C. And finally, a few people claimed to have seen a green rainbow following Salman al-Malahi, which, according to them emerged from Room C. All these incidents lead people to believe that the room contained some kind of a holy essence, of a type usually emanated by holy tombs. The conclusion was that most probably the house was built unknowingly on the location of a forgotten tomb that should, therefore, be rebuilt. As a result the tomb in room C was constructed, commemorating the unknown saint who supposedly was buried earlier in that very place.

The family lived in the house until the 1948 war, after which they were relocated to Bir Kuhle, 10 km to the north of Bet al-Malahi. Later the family moved again and finally settled down in Rahat (*ca.* 5 km north of Bet al-Malahi). Currently the Bet al-Malahi complex serves as a

[1] The family tradition claims that the al-Malahi originated as a branch of the bedouin tribe of Al-Harb who had migrated from Arabia to Sinai 600 years ago. In a later phase, some of them arrived in the Judean Desert escaping a blood vengeance, and later on (probably at the end of the 19th or beginning of the 20th centuries) finally reached the Northern Negev (Al-Turi 1996).

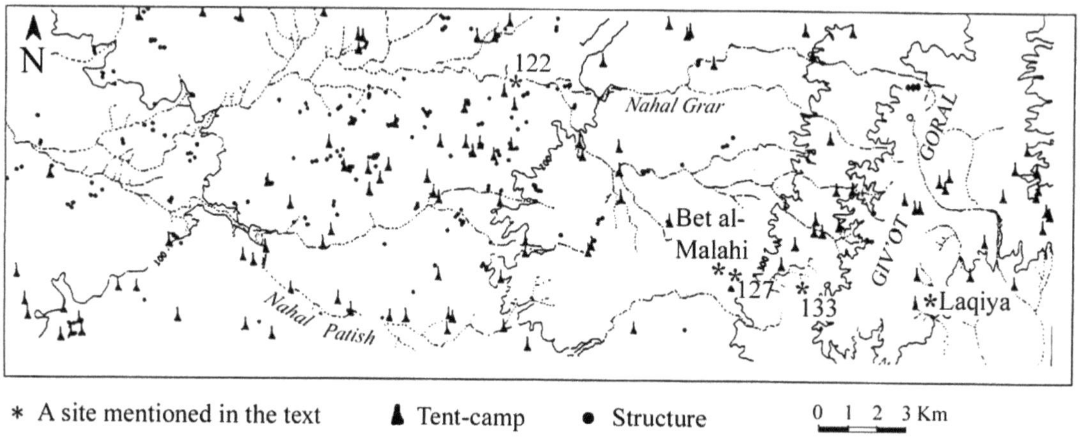

Fig. 9. Map of the studied area with indication of structures' and tent camps' locations in 1945 (according to aerial photographs)

place for local pilgrimage, which is focused on the tomb in room C. This is the reason for the marked path from the courtyard's entrance to the room. Pilgrims usually bring a sacrifice, which they slaughter outside the structure. The ceremonial cooking and eating take place in the courtyard. Visitors place incense (*Bahura*) on the tomb and next to it, as well as tin cans and small bottles with olive oil. They add a linen wick to it and light this simple lamp on the tomb. In addition small pieces of cloth are put on the tomb, and white flags are hung on the walls of the structure, and especially on the walls of Room C.

Tents, Solid Structures and Aerial Photographs

In order to complement the archaeological inquiry and the ethnographic interview we have studied two sets of aerial photographs of the region, from 1945 and 1949. In order to gain a wider perspective of the region, the examined area was larger than the immediate vicinity of Bet al-Malahi. The study area therefore constitutes a rectangle of 11 by 36 km along the upper basin of Nahal Grar, from Laqiya in the east to the confluence of Nahal Grar and Nahal Patish in the west (Figs 2, 9). The total area included in this inquiry thus measures 396 sq km.

The photographs indicate that in 1945 the inhabitants of the study area lived both in tents and in solid structures, such as Bet al- Malahi. In contrast, the 1949 set shows only solid structures (quite probably deserted) in the region and no tents – a result of the relocation of the bedouin community following the 1948 war.

Analysis of the 1945 aerial photographs enabled us to map all the recognizable tents, tent camps and solid structures in the vicinity of Bet al-Malahi (Figs 2, 9; Yekutieli and Gabai 1995). In addition to this general mapping, specific features of each dwelling type were studied. For example, the photographs indicate that bedouin encampments were positioned in one or two parallel rows, on a north-south axis, and that most of the campsites were located on the lower part of slopes above the *wadi* beds.

Turning to the solid structures, the images reveal that they belonged to two main categories: simple buildings of a square or a rectangular plan, and courtyard structures consisting of an open courtyard delimited by a high perimeter wall, and one or more flanks of roofed area comprising typically between four to eight rooms. It appears that the majority of solid structures in the region were simple buildings (85% – 178 structures), while courtyard structures were the minority (15% – 32 structures).

The aerial photographs further reveal variations in the clustering patterns of tents versus solid structures. For example, the tents tended to cluster together in camps, the largest of which were located in the area of present day Laqiya. Here one camp comprising 21 tents was located, one km removed from a second camp of 20 tents. In contrast to the tents, solid structures usually did not cluster together. As a rule, the largest concentration of solid structures in one location did not exceed seven buildings.

Looking at the overall numbers, the examination reveals that prior to the 1948 war there were 210 solid structures in the region, and 118 tent-camps with a total number of 605 individual tents. Thus the region's population lived concurrently in solid and temporary structures in a ratio of 1:3.

The aerial photographs indicate too that the distribution of solid versus temporary architecture had a clear pattern: tents are the predominant dwellings in the eastern and southern parts of the studied area, and in contrast, solid structures are more numerous in its western and northern sections. A complementary observation relates to cultivated versus uncultivated areas, a feature that is clearly visible in aerial photographs. Apparently the western 60% of the study area were ploughed and cultivated while the eastern 40% were not. This observation suggests a direct (although not exclusive) link between farmed area and solid architecture on the one hand, and uncultivated land and tent camps on the other.

The reasons for the latter phenomenon might be sought in the topographic and climatic conditions of the study area described above: Apparently, within the research area regions with rough topography, as in its eastern part, and regions with less rainfall such as its southern part, were more likely to be occupied by tents. In contrast, those regions with gentler topography and a better precipitation seem to be linked with solid architecture. In other words, in the area under study regions favourable for field agriculture were preferred for solid structures, while areas suitable for grazing were dominated by tents' scenery.

Zooming in on Bet al-Malahi, the aerial photographs reveal that it was one of the easternmost solid structures in the research area (Figs 9,10). It was set within a cultivated landscape that included a terraced wadi bed with an enclosed area (whose remains were still visible in 1993) situated immediately to its north. Several small tent camps were discerned in the vicinity of Bet al-Malahi, the closest of which (site 127) was half a kilometre to its southeast.

Fig. 10. A 1945 aerial photograph of Bet al-Malahi and its immediate vicinity (north on top)

Discussion

The investigation of Bet al-Malahi and its surroundings using various methodologies has unveiled its history as outlined above. In addition, this case study enables us to reflect upon certain common notions relating to the archaeology of ancient near eastern fringe zones. In the following pages I wish to focus on two such issues: courtyard plan interpretations, and the alleged lack of archaeological "finger-print" of recent bedouin populations in the Negev.

The Courtyard Structure plan

Since Bet al-Malahi is a courtyard structure, its study led us to examine the archaeological literature that deals with this architectural plan in our region. As this literature suggests, courtyard structures, whether square, circular or irregular in shape, are common throughout the history of the southern Levant. They appear in the region from the Late Pre-Pottery Neolithic B at the latest (Nissen *et al.* 1991), and continue to exist until the present (e.g. Chalcolithic courtyard structures: Porat 1987: 38-19; Bronze Age: Beit Arieh 1987: 72; Iron Age: Gophna 1963; Finkelstein 1985; Hellenistic-Roman: Hirschfeld 1987: 85).

A special interest in the plan of the courtyard structure is perceptible in publications of the biblical archaeology genre of recent decades. Fritz and Kempinski suggested that the so-called four-room structure, typical of the Iron Age, is a specific case of a courtyard structure (1983: 31). They described it as a central open-air hall flanked by roofed rooms on three sides, and suggested that its origin was the nomads' tent. In their Tel Masos excavation report they present a plate in which they propose an evolution from a rectangular bedouin-like tent with an activity area in front of it, to a fully developed four-room structure, in which the former activity area evolved into a central open-air aisle. The concept they wished to convey through this evolutionary model was that the Iron Age I (1,200-1,000 BC) four-room house represented a final stage in a process of settlement of previously tent dwelling nomadic groups. In the specific context of their discussion they were referring to a clear historical issue: Israelite and/or related tribes who, according to the Bible (apparently followed by Fritz and Kempinski when they wrote the Tel Masos report), ceased to wander and settled down. It should be noted that the idea raised some criticism (e.g. Stager 1985: 17; Netzer 1987: 167, note 7), and apparently, although Kempinski probably continued to hold this view, Fritz had given it up in 1995 (Fritz 1995: 74, note 33).

The interest of Biblical Archaeology in the courtyard plan was not confined to the four-room house. Israel Finkelstein developed a comparable evolutionary model to that of Fritz and Kempinski, focusing on a different, although possibly related phenomenon: that of the Negev Highlands "forts" (some of which are in effect in the same size as Bet al-Malahi; Cohen 1995, fig. 17), and the oval shaped courtyard village plans of sites like 'Izbet Sartah (Finkelstein 1986:218-228). On the basis of two photographs made by early 20th century travellers depicting bedouin tents arranged around an open courtyard (Finkelstein 1986: 225, figs. 80, 81), Finkelstein suggested that the stone built Iron I courtyard sites are the solid execution of that arrangement of tents. Thus, he proposed, the courtyard sites are manifestations of settling nomads, either the proto-Israelites or a closely related group.

Our research of Bet el-Malahi and its cultural environment leads us to join those who object this proposed evolutionary link between nomadic architecture and solidly built courtyard structures and sites.

Firstly it seems that the evolutionary line drawn between a bedouin-tent and a four-room structure is a purely theoretical experiment with no empirical support. In other words, in theory we can draw plates that will "prove" that any kind of structure can be transformed in a few stages to any other existing plan.

Secondly our case sheds some doubt on the inclusive validity of Finkelstein's hypothesis that courtyard sites are a direct manifestation of settling nomads' presumed translation of the tent camp into solid architecture. The case we have described above is in effect a case of settling nomads. Following Finkelstein's line of thought it would be expected that in this case too, where courtyard structures "suddenly" emerged throughout the region, they would reflect existing tent-camps' formations. However, the 118 tent-camps (with a total number of 605 tents), which we have observed are all arranged in rows, and not a single one displays an enclosed pattern. Nor do the groupings of solid structures reflect either an arrangement in two rows, or a 'courtyard' formation. Thus, settling nomads are not necessarily reconstructing their tent camps in stone or bricks.

Thirdly, at least in the case of the Al-Turi family, a village master builder built their courtyard structure. We cannot guess if the plan was ordered by the bedouin family or suggested to them by the master builder. However it clearly reflects an ongoing interaction between sedentary and non-sedentary communities.

Finally, if we look for parallels for the courtyard structure plan it appears that it is a common rural and urban building tradition (Dever 1992: 52), prevalent since antiquity from Anatolia in the east to the Maghreb in the west (if not even in a wider territory). This plan apparently represents a set of ideas, mainly related to privacy (Insoll 2003: 19), moral values and climate control.

Thus, our conclusion is that it is much more reasonable to say that the appearance of courtyard structures amongst nomad societies of the southern Levant, including our case study of Bet al-Malahi, is an aspect of their contact and inclusion within the wider Mediterranean society and lifestyle, rather than a manifestation of a nomadic architectural concept built in stone.

We further believe that the insistence of some biblical archaeologists that the courtyard house plan derives from nomad architectural conceptions represents a dialectic relationship to their underlying assumption that the Israelite and related peoples appearing on the southern Levantine scene at the 13th century BC were settling nomads.

The Archaeological "Finger-Print" of recent bedouin Populations in the Negev

The issue of the courtyard house is linked to another theory about the relations between the nomads and the settled population (fully developed in Finkelstein 1995). A basic assumption in this theory is that archaeologists cannot identify "certain kinds of nomads", and therefore absence of archaeological remains in a given territory does not imply absence of large-scale human activity in the same area.[2] In Finkelstein's words:

> "The major questions to be confronted are the archaeological representation of nomads and our ability as archaeologists to track down all forms of human activity in desert environment – sedentary and non-sedentary alike. More specifically, can we make pastoral people visible?" (Finkelstein 1995: 23).

In order to strengthen his theory that "some nomads" are invisible Finkelstein claims that recent bedouin populations in the desert regions of the southern Levant leave few archaeological remains:

> "The most convincing example to the elimination of the theory "No remains, no human activity" is the bedouin population of the Negev during the last centuries. Sources from the end of the Ottoman period and the British Mandate show that in the end of the last century and the beginning of the current one, there were between 8,500 to 16,500 people in the Negev Highlands. However the only hints for their presence are cemeteries and tent-camp remains, many of which are washed away through the centuries. The same is true in regard of the bedouin population in large parts of Sinai" (my translation of Finkelstein 1989: 76).

Before wondering about the validity of this theory in light of our research, a short historiographic voyage is illuminating since it portrays how this claim was intercepted in the literature after Finkelstein's initial statement was published (each reference is preceded by the year of its publication):

1989 – "[...] However the only hints for their presence are cemeteries and tent-camp remains, many of which are washed away through the centuries." (my translation of Finkelstein 1989:76).

1995 – " [...] except for the Hajj forts, the northern Sinai road and sites in the vicinity of St. Catharine's monastery

[2] This was an ongoing debate in the 1990s; there were those who claimed that archaeology is unable to discover the faint remains that nomads leave behind (Finkelstein and Perevolotsky 1990; Finkelstein 1995), while others insisted that a detailed archaeological research would identify them (Rosen 1992).

no material remains from the long period between the 8th century and the late Ottoman period have been found in the Negev and Sinai" (Finkelstein 1995: 30).
1997 – "Finkelstein [...] documents several examples of peoples who are mentioned in historical/literary documents from ancient times YET who left NO trace (archaeologically) of their existence in the field (!): [...] even bedouin tribes in the first part of the 20th century (known from modern sources)" (Miller 1997; Emphasized in the original)
2001 – "As for the issue of encampments, it is nearly impossible to find traces of large bedouin encampments in the Sinai Desert from 200-300 years ago. So would one expect the remains of large encampments after 3,000 years?" (Lichtman 2001).
2003 – "In Palestine under the British Mandate (i.e. during the first half of the present century) between 55,000 and 65,000 bedouin lived in the Negev. I. Finkelstein comments: 'This population left almost no material remains, however; without contemporary, documentary evidence, we would not know of its existence' [1986: 51]. We should not expect semi-nomadic Israelites to have been any different in this respect." (Bimson 2003).

The clear point of this very brief survey is that Finkelstein's first observation (1989: 76; cited above), the only one in this string that stems from empirical field-work, and that actually admitted that nomad bedouin populations do leave archaeological evidence ("cemeteries and tent-camp remains"), was transformed by a chain of writers (including Finkelstein himself), into a clear-cut "fact" that thousands of bedouin mentioned in recent historical documents are invisible in the field.

Ostensibly that "fact" became a cornerstone for some writers, in solving the mystery of why the remains of hundreds of thousands of Israelites, which the Bible reports to have wandered in the desert, are not found: if recent nomadic bedouin are invisible, can we expect 3,000 years old Israelite nomads to be visible?
Thus a basic archaeological claim has turned into a part of an ideology that establishes its own expectations and assumptions. Under these circumstances it will be more useful to study this issue using the methodology of discourse analysis (Barker and Galasinski 2001), rather than archaeological investigation.
However, until this is done it is useful to return to another of Finkelstein's recommendations:

> "...adherents of the 'no remains no human activity' model can save themselves time and frustration – until more sophisticated methods are invented, more exhaustive fieldwork will not solve the problem of the invisible nomads, the comfortable library would" (Finkelstein 1995: 30).

Our research of Bet al-Mala<u>h</u>i and its surroundings might provide such a "sophisticated method" which is practiced through a combination of work in the "comfortable library" and in the indispensable field. The 1945 aerial photographs database clearly indicates the locations and sizes of tents and tent camps. Thus a visit to those places could reveal whether they left archaeological remains or not. Accordingly a sample of tent camps in the vicinity of Bet al-Mala<u>h</u>i was selected for study in 1993.

We started at site 133 (Figs 2, 9). The 1945 aerial photograph indicates that it consisted of 13 tents located along a small *wadi* (Fig. 11). Our visit to the site proved that it is rich in remains: tens of Gaza Ware sherds, metal pieces, hundreds of baked stones (grey coloured limestone chunks from old hearths), some undefined stone concentrations, and even an occasional World War I shell (Figs 12-14). Additionally, a *wadi* that had recently cut through the site revealed some ashy patches, probably the remains of hearths within the tent camp.

Notably site 133 was a large camp situated within an area that was not disturbed since its desertion at the end of the 1940s. Therefore, in order to further assess our identification capabilities, a much smaller camp in a more disturbed area was chosen. With the help of the aerial photographs, site 127 was selected. This site, which consisted of three tents in 1945 (Fig. 15), is currently situated within a sparsely wooded area, 400 m south of Bet al-Mala<u>h</u>i and two km west of site 133 (Figs 2, 9). It took one surveyor twenty minutes of work on the site to reveal pieces of Gaza Ware as well as many heat-baked greyish stones (Fig. 16).

Fig. 11. A 1945 aerial photograph of site 133 (north on top)

Fig. 12. Site 133 in 1993, looking east

Fig. 13. A WWI shell found at site 133

Fig. 14. Collection of sherds and metal pieces from site 133

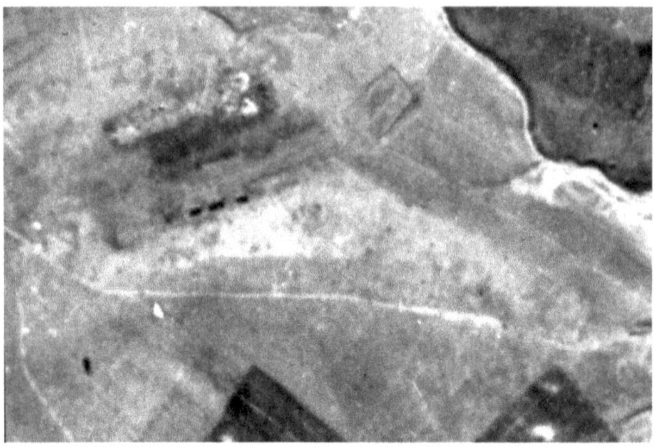

Fig. 15. A 1945 aerial photograph of site 127 (north on top)

Fig. 16. Collection of sherds and burnt hearthstones from site 127

Another tent-camp (site 122; Figs 2, 9) currently located in an extremely disturbed area, was also selected. The location where it lied in 1945 is currently an intensively ploughed and irrigated field that has been tilled by members of kibbutz Mishmar Hanegev from the late 1940s onwards. At the site's former location a small number of remains were found, most of which were heat-baked hearth stones.

Accordingly, our first conclusion is that tent sites are identifiable, though special attention should be given even to small numbers of sherds and to the usually unnoticed phenomenon of hearthstones. Secondly, our study, which focused on 1945 tent-camps, allows the conclusion that the idea that bedouin of the British mandate period left no archaeological fingerprint is incorrect.

These two conclusions call for a continued fieldwork on many new trajectories. For example, it is time to return to the numerous archaeological remains of tent-camp sites that were detected in the field, but did not show up on the aerial photographs, and attempt to date them. This can be done either by systematic carbon dating of their hearths, or by studying their Gaza Ware collections, which apparently existed in the region at least from the 17th century onwards (Rosen and Goodfriend 1993).

Conclusions

The study described above has led to a better understanding of Bet al-Mala$\underline{h}$i and its environment in the 1940s, and to a few theoretical conclusions.

It became clear that, in the 1940s, the area under study witnessed a coexistence of tent and solid structures' dwellers that may have belonged to the same families. These people relied on mixed subsistence patterns that among other things included field agriculture and herding. The last two activities were apparently dependent on topographic and climatic conditions, and interconnected with specific architectural patterns: more solid structures were observed in the agricultural zone, and tents were more frequent in the grazing area.

Focusing on Bet al-Malahi, this study has proven (once again) that the archaeological method may identify in a precise way details such as building phases within a structure, but that it is much weaker in supplying social interpretations. Without 'Id al-Turi's evidence we would never have known that the tomb in room C was built deliberately without any interment (Note a similar, contemporaneous case described by Canaan 1924: 54-55). If it had been excavated we would probably have proposed, as archaeologists often tend to do with empty tombs, that the bones disintegrated, or were collected and taken somewhere else, or just thrown away when the tomb was robbed (i.e. Epstein 1975: 1). Similarly there was no way we would have known that the stone piles south of the building were designated for the building of a school, and that the pits on its eastern side were meant for the storage or hiding of grain.

The short study of the courtyard plan, which followed our fieldwork, clarified that it is not necessarily a nomadic invention, and that it is rather a manifestation of the interaction between settled and nomadic communities of the southern Levant. Under these conditions the question whether the settled community borrowed the plan from the nomads or vice versa becomes pointless.

Lastly the idea that bedouin tribes of the 18th to 20th centuries left no archaeological fingerprint in the Negev has been disputed, and it has been suggested to direct our efforts towards a more rigorous dating of the numerous nomad tent camps found in archaeological surveys in the fringe zones.

Bibliography

Palestine 1:20,000 Map (1947). Jaffa: Survey of Palestine.

Abu-Rabi'a, A.
2001 *A Bedouin Century*. New York: Berghahn.

Al-Turi, I'.
1996 "The History of the Al-Malahi Tribe in Al-Turi, 'I'". *The Animal Kingdom*. Jerusalem: Ariel. (Hebrew).

Barker, C., and Galasinski, D.
2001 *Cultural Studies and Discourse Analysis: A dialogue on language and identity*, London: Sage.

Barslavski, Y.
1947 *Hyada'ata et Ha'aretz. Vol. B. The Negev*. Tel Aviv: Hakibutz Hameuchad. (Hebrew).

Beit Arieh, I.
1987 "Construction and Settlement Pattern in Southern Israel and Southern Sinai During the Early Bronze Age II" in *The Architecture of Ancient Israel*, eds H. Katzenstein; E. Netzer; A. Kempinski; and R. Reich. Jerusalem: Israel Exploration Society. (Hebrew). pp. 71-73.

Bimson, J.J.
2003 Exodus and Conquest – Myth or Reality? Can Archaeology Provide the Answer? *Journal for the Ancient Chronology Forum* 2 (to be found on: http://www.ouviroevento.hpg.ig.com.br/textos/arqueologiaehist/johnjbimsonexodus.htm last visited september 19, 2005.

Canaan, T.
1924 "Mohammedan Saints and Sanctuaries in Palestine". *Journal of Palestine Oriental Society* 4: 1-84.

Cohen, R.
1995 "Fortresses and Roads in the Negev during the First Temple Period" in *Eilat: Studies in the Archaeology, History and Geography of Eilat and the Arava*, eds J. Aviram; H. Geva; R. Cohen; Z. Meshel; and E. Stern. Jerusalem: The Israel Exploration Society and Israel Antiquities Authority. (Hebrew). pp. 80-126.

Dan, Y.
1988 "The Soils of the Southern Shefelah" in *Man and Environment in the Southern Shefelah*, eds D. Urman, and E. Stern. Tel-Aviv: Massads (Hebrew). pp. 50-58.

Danin, A.
1988 "Vegetation in the Southern Shefelah" in *Man and Environment in the Southern Shefelah*, eds D. Urman, and E. Stern. Tel-Aviv: Massads (Hebrew). pp. 59-64.

Dever, W.G.
1992 "How to Tell a Canaanite from an Israelite" in *The Rise of Ancient Israel*, eds H. Shanks; W.G. Dever; B. Halpern; and P. Kyle McCarter. Washington DC: Biblical Archaeology Society. pp. 27-60.

Epstein, C.
1975 "The Problem of the Golan Dolmens According to the Latest Excavations. *Eretz-Israel* 12: 1-8. (Hebrew).

Finkelstein, I.
1985 "The Negev Highland 'Fortresses' in the Iron Age: Settlement Sites of the Desert Nomads". *Eretz-Israel* 18:366-79 (Hebrew).
1986 *The Archaeology of the Period of Settlement and Judges*. Tel Aviv: Hakibutz Hameuchad. (Hebrew).
1989 "The Economic and Demographic Structure of the Intermediate Bronze Age: A Reassessment". *Eretz-Israel* 20:75-81 (Hebrew).
1995 *Living on the Fringe*. Sheffield: Sheffield Academic Press.

Finkelstein, I., and Perevolotsky, A.
1990 "Processes of Sedentarization and Nomadization in the History of Sinai and the Negev". *Bulletin of the American Schools of Oriental Research* 279: 67-88.

Fritz, V.
1995 *The City in Ancient Israel*. Sheffield: Sheffield Academic Press.

Fritz, V., and Kempinski, A.
1983 *Ergebnisse der Ausgrabungen auf der Hirbet el-Msas (Tel Masos), 1972-1975*. Wiesbaden: Harrassowitz.

Gophna, R.
1963 "Hatzerim in the Northern Negev". *Yedio't* 3: 173-180. (Hebrew)

Herzog, Z.
1990 "From Nomadism to Monarchy in the Beer-Sheba Valley" in *From Nomadism to Monarchy*, eds N. Na'aman, and I. Finkelstein. Jerusalem: Yad Izhak Ben-Zvi and The Israel Exploration Society. (Hebrew). pp. 215-41.

Hirschfeld, Y.
1987 *The Palestinian Dwelling in the Roman-Byzantine Period.* Jerusalem: Yad Izhak Ben-Zvi. (Hebrew).

Insoll, T.
2003 *The Archaeology of Islam in Sub-Saharan Africa.* Cambridge: Cambridge University.

Lichtman, D.
2001 Archaeology and the Bible – part 2. http://www.aish.com/societyWork/sciencenature/Archaeology_and_the_Bible_-_Part_2.asp, (last accessed September 19, 2005).

Miller, G.
1997 http://www.christian-thinktank.com/noai.html (last accessed September 19, 2005).

Netzer, E.
1987 "Dwelling Structures of the Iron Age" in *The Architecture of Ancient Israel*, eds H. Katzenstein; E. Netzer; A. Kempinski; and R. Reich. Jerusalem: Israel Exploration Society. (Hebrew). pp. 165-172.

Nissen, H.; Muheisen, M.; and Gebel, H.G.
1991 "Report on the Excavations at Basta 1988". *Annual of the Department of Antiquities of Jordan* 35:13-40.

Porat, Y.
1987 "Dwellings of the Chalcolithic Period" in *The Architecture of Ancient Israel*, eds H. Katzenstein; E. Netzer; A. Kempinski; and R. Reich. Jerusalem: Israel Exploration Society. (Hebrew). pp. 37-44.

Rosen, S.A.
1992 "A Response to Finkelstein and Perevolotsky". *Bulletin of the American Schools of Oriental Research* 287: 75-85.

Rosen, S. A., and Goodfriend, G. A.
1993 "An Early Date for Gaza Ware from the Northern Negev". *Palestine Exploration Quarterly* 125:143-148.

Stager, L.W.
1985 "The Archaeology of the Family in Early Israel". *Bulletin of the American Schools of Oriental Research* 260: 1-35.

Yekutieli, Y. and Gabai, M.
1995 "Survey of Beduin Structures in the Northern Negev in the 1940s – A Test Study of the Archaeology of Nomads". *Archeologia* 4:43-57 (in Hebrew).

Zangvil, A.
1988 " The Climate of the Southern Shefelah" in *Man and Environment in the Southern Shefelah*, eds D. Urman, and E. Stern. Tel-Aviv: Massads (Hebrew). pp. 42-49.

www.ingramcontent.com/pod-product-compliance
Lightning Source LLC
LaVergne TN
LVHW071218210926
848482LV00015B/844